Traditional and Complementary Medicine

Edited by Cengiz Mordeniz

Published in London, United Kingdom

IntechOpen

Supporting open minds since 2005

Traditional and Complementary Medicine
http://dx.doi.org/10.5772/intechopen.78452
Edited by Cengiz Mordeniz

Contributors
Silvio Leite Monteiro Da Silva, Mmamosheledi Mothibe, Mncengeli Sibanda, Arun Rasheed, Cengiz Mordeniz

First published in London, United Kingdom, 2019 by IntechOpen
IntechOpen is the global imprint of INTECHOPEN LIMITED, registered in England and Wales, registration number: 11086078, 7th floor, 10 Lower Thames Street, London, EC3R 6AF, United Kingdom
Printed in Croatia

British Library Cataloguing-in-Publication Data
A catalogue record for this book is available from the British Library

Additional hard and PDF copies can be obtained from orders@intechopen.com

Traditional and Complementary Medicine
Edited by Cengiz Mordeniz
p. cm.
Print ISBN 978-1-78984-183-1
Online ISBN 978-1-78985-377-3
eBook (PDF) ISBN 978-1-78985-378-0

For EU product safety concerns:
IN TECH d.o.o., Prolaz Marije Krucifikse Kozulić 3, 51000 Rijeka, Croatia, info@intechopen.com or visit our website at intechopen.com.

Meet the editor

Cengiz Mordeniz, Associate Professor at Tekirdag Namik Kemal University, is the Head of Pain and Traditional and Complementary Medicine Clinics. Previously, he held academic positions at Dicle and Harran Universities. He has visited Rigs Hospitalet, Denmark; Heidelberg University and Giessen University, Germany; and Plovdiv University, Bulgaria. He was trained at the Moscow Quantum Medicine Academy and School of Advanced International Studies on Applied Theoretical and Non-Linear Methodologies of Physics, Italy. He participated in the Harvard University Clinical Research Program. Professor Mordeniz completed his elementary education at Galatasaray High School, Ankara School of Science, and Tate High School, USA. He earned an MD and Anesthesiology and Intensive Care specialization from Istanbul Medical Faculty and Master of Medical Law and Clinical Deontology at Acibadem University.

Contents

Preface **XI**

Chapter 1 **1**
Introductory Chapter: Traditional and Complementary Medicine
by Cengiz Mordeniz

Chapter 2 **23**
Integration of Traditional and Complementary Medicine into Evidence-Based Clinical Practice
by Cengiz Mordeniz

Chapter 3 **31**
African Traditional Medicine: South African Perspective
by Mmamosheledi E. Mothibe and Mncengeli Sibanda

Chapter 4 **59**
La Gioconda Effect and Self-Organizing Factors of Biofield
by Silvio Leite Monteiro da Silva

Chapter 5 **75**
A Review on Natural Antioxidants
by Arun Rasheed and Rinshana Fathima Abdul Azeez

Preface

Over the past century, the chemical industry has facilitated the rapid growth of the pharmaceutical and food industries: from the milk we drink to the medicines we take, chemicals have had a remarkable impact on the wellbeing of human beings. Moreover, advances in biotechnology, epigenetics, and medical sciences have been exploited for the sake of privatization and commercialization of the health sector. The focus shifted from treating the disease to having a cost-effective solution to alleviate the symptoms. Hence, patients were seen as customers and physicians as professional health service providers.

Doctors began to follow a standard algorithm for the diagnosis and remedy of the signs of the patients instead of focusing on human interaction. Exacerbated by the challenges imposed by insurance companies and legal framework, punishment against malpractice also pushed doctors into defensive medicine. Ultimately, the biopsychosocial/spiritual aspect of the human being has been dismissed, and the standardized service has been considered as a successful practice.

Scientific research has proven the placebo and nocebo effects of drugs and therapeutic methods. The satisfaction of patients is evaluated by oral or written inquiries, which may not always reflect their true feelings and wellbeing. Patients have simply become a virtual identity with their laboratory findings and images; they are immediately diagnosed with a disease or syndrome based on those results and put into a pharmacological therapy, which may continue for a lifetime for those with chronic diseases. Moreover, new drugs are ordered frequently to compensate for complications and side effects. It is inevitable that physicians may lose their flexibility for taking initiatives and experimenting with novel solutions. This mechanical structure, consisting of orthodox doctors and customer patients who have to give written consent to establish trust, is soon to be replaced by personalized and holistic treatments.

Recently, traditional medicine and methods have regained popularity, promising a more holistic, organic, and personal treatment. However, unregulated applications can potentially cause risks since any medicine or method could trigger side effects or complications. Although the World Health Organization, European Union, and some official authorities have started new regulations and training programs, additional research and evidence-based literature are required to train health professionals to apply traditional medicine.

This book is written in the hopes of inspiring and motivating people who are interested in learning more about some of the oldest medical practices and what the future holds for them. The introductory chapter reviews in general the phytotherapy, homeopathy, biofield therapies, and general traditional medicine emphasizing African traditional medicine. Other chapters explain one of the oldest traditions that still survives in Africa and illuminate on the physical aspect of the human being as a biofield. Yet another chapter is related to chemical aspects, expounding on oxidative stress and antioxidant natural products. And a further chapter evaluates evidence-based aspects of traditional medicine.

We hope that this book takes a valuable place in the integration of traditional medicine with evidence-based medicine.

Cengiz Mordeniz
Namik Kemal University,
Tekirdag, Turkey

Chapter 1

Introductory Chapter: Traditional and Complementary Medicine

Cengiz Mordeniz

1. Phytotherapy

Plants are man's first medicines. In ancient times, "medical" care mainly involved using plants as medicines. Throughout history, people all over the world have used herbs to maintain and improve health into the art of healing. A constant process of searching, testing, and verifying in all cultures across the globe resulted in the development of an empirical science. Many plants have an established place within scientific medicine and are used for a broad range of health conditions. A herb is a plant or its part used for its scent, flavor, or therapeutic properties. They are sold as tablets, capsules, powders, teas, extracts, and fresh or dried plants.

Herbal medicine as the oldest form of health care is the synthesis of therapeutic experiences of generations for over hundreds of years and has gained popularity again in today's medical practice with over 85% of the world's population using phytotherapeutic medicines according to WHO [1].

The increase in the use of herbal products is due to their cultural acceptability, availability, affordability, efficacy, and safety claims. Primitive humans distinguished useful herbs with beneficial effects from those that were inactive or toxic. The basis of modern medicinal drugs such as aspirin, morphine, digitoxin, and quinine is through scientific validation of herbal medicine. Review of different national pharmacopeia reveals that at least 120 distinct chemical products/moieties from herbal sources have been utilized as lifesaving drugs.

In 2010, the inventory of a British-American team of researchers revised the previously stated number of plants down from 900,000 plants to between 300,000 and 400,000 of which only 6% have been screened systematically for their biological activity and 15% have been investigated phytochemically and just under 10% have had any form of research carried out into the possible use as medicines. It is predictable that natural compounds and their derivatives comprise nearly 60% of all drugs in clinical use and medicinal plants contribute not less than 25% [2].

In medical systems based on herbalism, folklore, or shamanism, written documents do not exist, and the herbal formulations are often kept secret by the practitioners, making the information more difficult to access [3].

The practices and the philosophy of each herbal medicinal system are influenced by their region: Ayurveda, a health care system that has been founded by ancient Hindu healers and saints, has been still used in India for over 5000 years. Its materia medica provides a comprehensive description of over 1500 herbs and 10,000 formulations. The Indian government has recognized Ayurveda to be a complete health care system in comparison with western Medicine and has compiled and preserved traditional medicinal knowledge in the public domain with The Traditional

Knowledge Digital Library (TKDL). Ayurgenomics aim to provide a base for human classification, diagnostics, and customized medicine with Ayurvedic concept of Prakriti from pharmacogenomics perspective, and AyuSoft has developed interactive software based on Ayurvedic classics as a decision support system. Herboprint uses three-dimensional HPLC to develop tools for activity-based standardization of botanicals.

A lot of medicinal plants, traditionally used for thousands of years, are present in a group of herbal preparations of the Indian traditional health care system named Rasayana proposed for their interesting antioxidant activities.

As metabolic diseases and age-related degenerative disorders are closely associated with oxidative processes in the body, the use of herbs and spices as a source of antioxidants to combat oxidation warrants further attention on validating the antioxidant capacity of herbs and spices after harvest, as well as testing their effects on markers of oxidation in parallel with clinical trials aiming to establish antioxidants as mediators of disease prevention. From a dietary perspective, the functionality of herbs and spices is exposed through consideration of their properties as foods. Through evidence-based frameworks for substantiating health claims related to foods, recommendations are warranted to support the consumption of foods rich in bioactive components, such as herbs and spices in the overall maintenance of health and protection from disease. In the book, a chapter is dedicated to the antioxidant properties of natural products [4–6].

In China, their own system known as the Traditional Chinese Medicine has been used throughout history. The oldest known herbal book in the world "The Devine Farmer's Classic of Herbalism" was compiled in China about 2000 years ago, including numerous herbal pharmacopeias and various monographs on specific herbs and their composition information.

The first textbook fully devoted to the description of herbal drugs is the Shen-nung-pen-ts'ao ching (ShenNung's Classic of Pharmaceutics) was written during the later Han period (25–220 AD). The literature of Chinese Materia Medica developed by continuous addition of new drugs as well as re-evaluation and addition of new indications for existing herbs during the centuries has been a valuable source for the acquisition of ethnopharmacological data and the development of new medicinal plants by studying the ancient textbooks.

The first compound derived from Chinese herbal remedies to enter the western market was ephedrine, an amphetamine-like stimulant from ma huang (*Ephedra sinica*).

Kampo medicine, the Japanese herbal medicine dates back over 1500 years with approximately 148 formulations [7].

In 1873, the Ebers Papyrus, the most ancient Egyptian medicine treaty—dated 1600 BC, was found, proving the use of plants for therapeutic purposes.

During the Trojan War (1200 AC), the plant *Achillea millefolium* was used on wounded soldiers in order to stop bleeding and heal wounds [8].

After the fall of the Roman Empire, the tradition of herbal medicine moved to the monasteries. The writings of famous healers from earlier times were copied, and many monasteries established and maintained herb and medicinal plant gardens, which led to the acquisition of new medico-botanical knowledge.

In Germany, herbal medicine is identified as one of the elements of naturopathy, and approximately 600–700 plant-derived medicines are accessible and prescribed by approximately 70% of German physicians. In 2011, 20% of herbal drugs were sold as prescriptions and 80% over the counter in Germany. In the EU, annual revenues from herbal medicines surpassed US$ 6 billion in 2003.

The National Canadian Institutes disbursed approximately 89 million dollars for research in traditional therapies in 2004.

In 2005, the National Centre for Complementary and Alternative Medicine at the National Institutes of Health in the USA spent about 33 million US dollars on herbal medicine.

In recent decades, the pharmacological properties of numerous medicinal plants and opportunities in phytotherapy have been explored through research projects, reviews, and monographs. These studies confirm that medicinal plants offer new approaches to tackling diseases. Herbal medicine has become a popular form of health care; even though several differences exist between herbal and conventional pharmacological treatments, herbal medicine needs to be tested for efficacy using conventional trial methodology. The public is often misled to believe that all natural treatments are inherently safe, but herbal medicines do carry risks. The triad absorption/metabolism/efficacy of herbs and their extracts is still an unsolved problem in judging their health effects. Some side effects such as allergic reactions, mutations, intoxication, teratogenesis, carcinogenesis, and medication interactions may occur if the use of the phytotherapeutic or medicinal plant is unrestricted. The lack of a stricter control of these medicines enables contamination by heavy metals, conventional drugs, herbicides or pesticides [9].

Based on World Health Organization reports, resistance of bacteria to antibiotics is a major global health challenge now and in the future. Different strategies such as inhibition of multidrug resistance pumps and biofilm formation in bacteria and development of new antibiotics with novel mechanism of action have been proposed to tackle this problem. Flavonoids, a large class of natural compounds, have been extensively studied for their antibacterial activity, in more than 150 articles published since 2005 and especially chalcones showed up to sixfold stronger antibacterial activities than standard drugs in the market. Some synthetic derivatives of flavonoids also exhibited remarkable antibacterial activities with 20- to 80-fold more potent activity than the standard drug against multidrug-resistant Gram-negative and Gram-positive bacteria (including *Escherichia coli*, *Pseudomonas aeruginosa*, and *Staphylococcus aureus*). Some of the flavonoids (i.e., quercetin) with a strong background of use in clinical trials are good candidates for further clinical studies as antibiotics alone or in combination with conventional antibiotics [10].

In the concept of the "doctrine of signatures," the healing effect of plants was deduced from their taste, shape, color, and other characteristics, for example, celandine (*Chelidonium majus* L.) being used as a remedy for the gallbladder and liver because of its yellow sap, orchids being used as an aphrodisiac because of their tubers resembling male testicles, and walnuts being used for mental illnesses because of their appearance resembling the surface of the brain.

The Nobel Prize in Physiology and Medicine 2015 was awarded for the discovery of two main natural products: (1) avermectin, a macrocyclic lactone isolated from the soil microorganism *Streptomyces avermitilis* (and its derivative ivermectin) and (2) artemisinin, a sesquiterpene lactone containing a peroxide bridge, isolated from the plant *Artemisia annua* L. (Asteraceae) [11].

The World Health Organization (WHO) defines herbal medicine as a practice, which includes herbs, herbal materials, herbal preparations, and finished herbal products, which contain as active ingredients parts of plants or other plant materials or combinations. These herbs are derived from plant parts such as leaves, stems, flowers, roots, and seeds.

Modern or scientific herbal medicine is also called phytomedicine or phytotherapy. Phytotherapy is a science-based medical practice with more traditional approaches, such as medical herbalism, which relies on an empirical appreciation of medicinal herbs linked to traditional knowledge.

Phytotherapy is defined as the study of the use of extracts of natural origin as medicines or health-promoting agents in an allopathic discipline, because they are directed against the causes and the symptoms of a disease. In Germany, phytotherapy is classified as a discipline of natural orthodox science-oriented medicine, with scientific requirements of the chemically defined substances in terms of quality, safety, and efficacy. Modern mass production of natural products as food supplements or herbal medicines may result in remedies that can differ greatly (dosage form, mode of administration, herbal medicinal ingredients, methods of preparation, and medical indications) from the traditions that form the basis for their safety and effectiveness, and an acceptable quality standard. The WHO has published monographs on the quality control, safety, and efficacy of selected medicinal plants and recommendations on their cultivation.

According to the World Health Organization, phytomedicine is defined as herbal preparations produced by subjecting plant materials to extraction, fractionation, purification, concentration, or other physical or biological processes. These preparations may be produced for immediate consumption or the basis for other herbal products. Such plant products may contain recipient or inert ingredients, in addition to the active ingredients.

Paraherbalism describes alternative and pseudoscientific practices of using unrefined plant or animal extracts as unproven medicines or health-promoting agents [12].

An herb may be any part of a plant including its leaves, stem, flowers, roots, and seeds. Herbal products may be raw or commercial preparations used to treat illnesses. Raw herbal products (leaves, seeds, or teas) are more used in less developed countries and commercial herbal preparations (tablets or pills) are more used in developed countries.

WHO published the requirements for clinical trials of herbal products, which contain the following definitions:

1. Herbal substance—material derived from the plant(s) by extraction, mechanical manipulation, or some other processes

2. Herbal product—the herbal material administered to clinical subjects

3. Herbal product synonyms—herbal remedy, herbal medicine, herbal drug, and botanical drug [13]

Herbal drugs, called phytochemicals, are the secondary metabolites of plants. Some are toxins, used to deter predation; some are pheromones, used to attract insects for pollination; others are phytoalexins, which protect against microbial infections; and yet others are allelochemicals, which inhibit rival plants competing for soil and light.

The first generations of plant medicine were simple botanical materials employed in more or less crude form. These medicines such as Cinchona, Opium, Belladonna, and Aloe were selected based on empirical evidence as gathered by traditional practitioners.

The second-generation phytopharmaceutical agents were pure molecules whose compounds differ from the synthetic therapeutic agent only in their origin, for example taxol from Taxus spp., quinine from Cinchona and reserpine from *Rauvolfia* spp.

In the development of third generation of plant medicine, the formulation is based on well-controlled double-blind clinical and toxicological studies with phytomedicine to improve the quality, efficacy, stability, and the safety of the preparations.

The following are characteristics of the phytomedicine:

- The active principle is frequently unknown.
- Standardization, stability, availability, and quality control are not easy.
- Well-controlled double-blind clinical and toxicological studies to prove their efficacy and safety are rare.
- They are cheaper than synthetic drugs.
- The belief that phytomedicine is devoid of side effect since millions of people all over the world have been using phytomedicine for thousands of years.
- The belief that phytomedicine is used for a wide range of treatment, especially of certain diseases where conventional medicine fails.

They are gentle, effective, and often specific in function to organs or systems of the body [14].

Although synthetic or chemical drugs can have greater or quicker effects than do equivalent phytomedicines, they present a higher degree of side effects and risks. For instance, psychopharmacological products with sedative and anxiolytic action are accompanied by undesirable side effects like uncoordinated motor skills and drowsiness, but phytomedicine acts on the body by regulating and balancing its vital processes rather than stopping or combatting certain symptoms. Its balancing effect prevents mental disorders and unbalanced mental condition.

The action of phytomedicines for the respiratory system is not limited to neutralizing the symptoms of any disease, but they also exert a true cleansing action for excessive mucus in the interior of the airway. They contain certain antibiotic substances that prevent bacteria growth in the mucus, for example *Thymus vulgaris* (thyme) and *Allium sativum* (garlic). Naringenin chalcone as the effective constituent of tomato extracts has been shown to inhibit the release of histamine from mast cells in the initial phase of inflammation and decrease the eosinophils and eosinophil cationic proteins by significantly improving nasal obstruction, rhinorrhea, and sneezing [15].

Phytomedicines are good dietary supplements, which are nutritive and replenish the body. For example, sunflower seed (*Helianthus annuus*) provides vitamin B6 (Pyridoxine).

Phytomedicines are effective in curing human pathogens like *E. coli* and *Candida albicans*, beyond symptomatic treatment of diseases and limit side effects associated with synthetic antimicrobial drugs. For example, *Hydrastis canadensis* not only has antimicrobial properties but also promotes optimal activity of the spleen in releasing compounds by increasing the blood flow in the spleen.

Recognition and application of phytomedicine depend on evidence-based clinical data. Phytomedicine can only enter in professional clinical application if safety, efficacy, and quality are proven in a comparable manner to conventional drugs. For this purpose, it is mandatory to conduct well-designed clinical trials. Awareness of Good Clinical Trial Practice and provision of knowledge worldwide is the existential foundation for proper scientific development [16].

"Medicinal plants" are neither phytomedicines nor phytotherapeutics. Once classified as a medicine, it is subject to the ethical standards defined by World Health Organization (WHO), and has to go through careful production processes from the time of collection and formulation until the time of packaging and distribution.

A plant may contain even thousands of chemical compounds acting in a synergistic way. Many conventional test methods are not always able to cope with the complexity of plant extracts to confirm the empirical and traditional use of a herb.

Many big pharmaceutical companies and scientists are returning to nature's apothecary in the search of new medicines [17].

Rational drug discovery from plants started when the German apothecary assistant Friedrich Sertürner isolated the analgesic and sleep-inducing agent from opium, which he named morphium (morphine) after the god of dreams, Morpheus, in 1805. This was followed in succession by many other herbal remedies or phytopharmaceutical substances (referred to in international terminology as HMPs or Herbal Medicinal Products). Strychnine from the poison nut tree (Strychnos nux vomica) in 1819, caffeine from the coffee bean (*Coffea*) in 1819, quinine from red cinchona (*Cinchona pubescens*) in 1820, codeine from opium in 1832, digitoxin from the purple foxglove (*Digitalis purpurea*), strophanthin from the seeds of the African climber (*Strophanthus gratus*), and atropine from deadly nightshade (*Belladonna atropa*) were important milestones in the discovery and isolation of plant-based constituents.

H.E. Merck in Darmstadt (Germany), the first apothecary extracting morphine and other alkaloids, was the first progenitor of pharmaceutical companies, which subsequently have produced natural products by chemical synthesis in order to facilitate production at higher quality and lower costs. Salicylic acid was the first natural compound produced by chemical synthesis in 1853.

The modern pharmaceutical industry laid its scientific and financial foundation after the discovery of penicillin (1928) from microbial sources. The therapeutic use of extracts and partly purified natural products has then been replaced by the use of pure compounds. For example, in the area of cancer, over the time frame from 1940s to the end of 2014, of the 175 small molecules approved 131, or 75%, are other than "S" (synthetic), with 85, or 49%, the actually being either natural or directly derived products.

Many big and medium-sized pharmaceutical companies have leaved their natural product research programs to academic universities and start-up companies [18].

Ethnobotany, as a research field of science, has been used for the documentation of indigenous knowledge on the use of plants for providing an inventory of useful plants from local floras. Plants that are used for traditional herbal medicine in different countries are an important part of the ethnobotanical studies, for the discovery of new drugs and new drug development. Over-harvesting, degradation of medical plants, and loss of traditional medical knowledge in local communities are common problems in the resource areas as well as issues of indigenous knowledge, intellectual property rights, and uncontrolled transboundary trade in medicinal plants [19].

Phytomedicine has become an important alternative treatment option for patients, as they seek to be treated in a holistic and natural way after an unsatisfactory response to conventional drugs.

Medical treatments cover the application of different components of plants (blossom, leaf, stem, and radix), aromatic essential oils, and herbal extracts as herbal teas, via massage as packs or wraps or in therapies using water, steam, or inhalation.

Although these subjects lost their importance in twentieth century because of the modern synthetic treatments, there is a renewed interest today in medicinal plants usage as natural products for the generation of semi-synthetic derivatives.

Currently, the paradigm of medicine has shifted from not only curing clinical diseases but also maintaining good health and enhancing quality of life with the integration of traditional medicine into the modern health care system [20].

2. Homeopathy

In conventional medicine, drugs are generally used to combat the symptoms of an illness. This principle is called allopathy (directed against an ailment) as well as for phytotherapy.

In homeopathy (similar to the ailment), a substance in diluted (potentized) form, which produces certain phenomena in healthy people, has a healing effect for a patient suffering from the same phenomena.

Common features of homeopathy and phytotherapy are that they emphasize a holistic approach and the regulation of self-healing powers and they are based on empirical values.

In determining the outcome of any traditional treatment, both in experimental and clinical settings including *forma mentis*, beliefs, knowledge, and practical abilities of the provider, as well as the positive or negative prejudices of the patient with respect to the provider of the therapy, cultural differences in the acceptability of the treatment and adherence to it, the patient-doctor encounter and differences in access to other treatments should be taken into consideration.

3. Traditional medicine and complementary and alternative medicine (TM/CAM)

The World Health Organization (WHO) distinguishes terminologically between traditional medicine and complementary and alternative medicine (TM/CAM). The WHO refers the term traditional medicine to developing countries (in Africa, Latin America, Southeast Asia, and/or the Western Pacific), as indigenous medicine (Traditional Chinese Medicine, Hindu Ayurveda, Arab Unani, and various forms of indigenous medicine) deeply rooted in history. The WHO uses the term CAM when referring to developed countries.

According to the World Health Organization (WHO), TM is "the sum total of the knowledge, skills and practices based on the theories, beliefs and experiences indigenous to different cultures, whether explicable or not, used in the maintenance of health, as well as in the prevention, diagnosis, improvement, or treatment of physical and mental illnesses, which may rely exclusively on past experience or observation handed down from generation to generation, verbally or in writing." It comprises therapeutic practices in existence for hundreds of years before the development of modern scientific medicine and is still in use today without much documented evidence of adverse effects [21].

TM therapies include medication therapies—use of herbal medicines, animal parts, and/or minerals—and nonmedication procedure-based therapies—without the use of medication, as in the case of acupuncture and related techniques, manual therapies, chiropractic, osteopathy, qigong, tai ji, yoga, naturopathy, thermal medicine, and other physical, mental, spiritual and mind-body therapies.

In countries where the dominant health care system is based on allopathic medicine, or where TM has not been incorporated into the national health care system, TM is termed as "complementary," "alternative," or "nonconventional" medicine." The term "traditional medicine" denotes the indigenous health traditions of the world; "complementary and alternative medicines" refer to methods outside the biomedical mainstream, as a set of health care practices that are not part of a country's own tradition and are not integrated into the dominant health care system, particularly in industrialized countries; and "conventional medicine" refers to "biomedicine" or modern medicine. Other terms that describe these health care practices, include "natural medicine," "nonconventional medicine," or "holistic medicine." The terms complementary/alternative/nonconventional medicine are used interchangeably with traditional medicine in some countries [22].

WHO's quality-of-life assessment of many traditional and complementary health systems includes spiritual dimensions of life and well-being, which is related to the sense of the meaning of the self or extending beyond the self. Expectancy based on belief and attitude causes a placebo, or "meaning response" effect on treatment outcomes in all therapeutic settings.

The traditional and complementary and alternative medicine (T/CAM) have claimed an increasing share of the public's awareness and the agenda of medical researchers. About half the population of many industrialized countries use T/CAM, (United States, 42%; Australia, 48%, France, 49%; Canada, 70%) and the proportion is as high as 80% in many developing countries (China, 40%; Chile, 71%; Colombia, 40%; up to 80% in African countries). Accompanied by a growth in research and associated literature, with an increase in an evidence-based approach, T/CAM has long been practiced both within and outside the dominant health care system. Most research has focused on clinical and experimental medicine (safety, efficacy, and mechanism of action) and regulatory issues, to the general neglect of public health dimensions. Public health research must consider social, cultural, political, and economic contexts to maximize the contribution of T/CAM to health care systems globally. A public health agenda is needed in addition to the focus on experimental research. Public health professionals need to define the public health dimensions of traditional and complementary medicines.

Practices of traditional medicine vary greatly from country to country, and from region to region, as they are influenced by factors such as culture, history, personal attitudes, and philosophy. In many cases, their theory and application are quite different from those of conventional medicine. The theories and concepts of prevention, diagnosis, improvement, and treatment of illness in traditional medicine (both herbal medicines and traditional procedure-based therapies) historically rely on a holistic approach toward the sick individual, and disturbances are treated on the physical, emotional, mental, spiritual, and environmental levels simultaneously taken into account the cultural background. As a result, most systems of traditional medicine may use herbal medicines or traditional procedure-based therapies along with certain behavioral rules promoting healthy diets and habits. Holism is a key element of all systems of traditional medicine.

Traditional medicine has not only continued to be used for primary health care of the poor in developing countries, but has also been used in countries where conventional medicine is predominant in the national health care system. Despite its existence and continued use over many centuries, and its popularity and extensive use during the last decade, traditional medicine has not been officially recognized in most countries. Long historical use of many practices of traditional medicine, including experience passed on from generation to generation, has demonstrated the safety and efficacy of traditional medicine. However, scientific research and evaluation of the traditional medicine are needed to provide additional evidence of its safety and efficacy, considering knowledge and experience obtained through the long history of established practices.

With the tremendous expansion in the use of traditional medicine worldwide, safety and efficacy as well as quality control of herbal medicines and traditional procedure-based therapies have become important concerns for both health authorities and the public. Various practices of traditional medicine have been developed in different cultures in different regions without a parallel development of international standards and appropriate methods for evaluating traditional medicine. Governments and researchers, among others, are increasingly requesting WHO to provide standards, technical guidance, and information to ensure that traditional medicine is used properly and to determine how research and evaluation of traditional medicine should be carried out.

Consequently, education, training, and research in this area have not been accorded due to attention and support. The quantity and quality of the safety and efficacy data on traditional medicine are far from sufficient to meet the criteria needed to support its use worldwide. The reasons for the lack of research data are due not only to health care policies but also to a lack of adequate or accepted research methodology for evaluating traditional medicine. It should also be noted that there are published and unpublished data on research in traditional medicine in various countries, but further research in safety and efficacy should be promoted, and the quality of the research should be improved.

Since 1991, WHO has developed and issued a series of technical guidelines such as guidelines for the assessment of herbal medicines, research guidelines for evaluating the safety and efficacy of herbal medicines, and guidelines for clinical research on acupuncture. However, these guidelines are still not sufficient to cover the many challenging issues in the research and evaluation of traditional medicine. In 1997, with the support of the National Center of Complementary and Alternative Medicine, National Institutes of Health, Bethesda, MD, USA, a WHO informal discussion developed draft guidelines for methodology on research and evaluation of traditional medicine.

Traditional medicine (TM), variously known also as ethno-medicine, folk medicine, native healing, or complementary and alternative medicine (CAM), is the oldest form of health care system as an ancient and culture-bound method of healing that humans have used to cope and deal with various diseases that have threatened their existence and survival. Different societies have evolved different forms of indigenous healing methods e.g., Chinese, Indian, and African traditional medicines. Traditional healer, on the other hand, is "a person who is recognized by the community where he or she lives as someone competent to provide health care by using plant, animal, and mineral substances and other methods based on social, cultural, and religious practices" [23].

4. Traditional medicine in Africa

Prior to the introduction of the cosmopolitan medicine, traditional medicine (TM) used to be the dominant medical system available to millions of people in Africa in both rural and urban communities as the only source of medical care for a greater proportion of the population.

The traditional health care systems are still in use by the majority of the people not only in Africa but across the world. In indigenous African communities, the traditional healers treat patients holistically by reconnecting their social and emotional equilibrium based on community rules and relationships unlike medical doctors who only treat diseases in patients. The arrival of the Europeans marked a significant turning point in the history of this age-long tradition and culture. During several centuries of conquest and invasion, European systems of medicine were introduced by colonizers and preexisting African systems were stigmatized and marginalized. Indigenous knowledge systems denied the chance to systematize and develop and even banned in some extreme cases. They were believed to be primitive and wrongly challenged by foreign religions dating back during the colonial rule in Africa and subsequently by the conventional or orthodox medical practitioners. In postindependence period, after a century of colonialism and cultural imperialism, TM has been recognized and held back the development of African traditional health care as an important aspect of health care delivery system in Africa. Despite the "passionate ambivalence," TM is still in use in modern day Africa after hundreds of years of its existence as a major African socio-cultural heritage without much reported cases of adverse effects.

The new health agenda in Africa focuses on the institutionalization of traditional medicine in parallel with orthodox medicine into the national health care scheme in order to move the health agenda forward since effective health cannot be achieved in Africa by orthodox medicine alone unless it has been complemented with traditional medicine.

As a whole, the annual market value of phytomedicine is close to $43 billion in all the world (more than some African annual budgets). Many African phytomedicines are well known in the international markets and Africa is one of the main world producers of the medicinal plants. For example, Cinchona yields quinine, an antiplasmodial drug for the treatment of malaria.

The WHO has helped most developing countries of the world by utilizing expert committees' policy decisions, and resolutions in providing guidelines that will aid the countries to develop and utilize their indigenous medicines for their national health agenda. In Africa, the health agenda is targeted at the recognition and development of phytomedicines by indigenous medical and pharmaceutical research scientists. This has lead the African Heads of States to declare the first 10 years of the millennium (2001–2010) as the "Decade of Traditional Medicine in Africa" and to celebrate on 31st August every year to make sure that phytomedicine is recognized and appreciated in the health sector.

WHO has offered a memorandum to help African member states in institutionalizing African traditional medicine in their health system and challenging different African Research centers on traditional medicine to cure priority diseases in Africa, such as malaria, HIV/AIDS, sickle-cell anemia, diabetes, and hypertension.

WHO has provided guidelines for institutionalization of traditional medicine into the health scheme including the below steps:

1. Political recognition: the government and heads of states should develop research on traditional medicine for the treatment of priority diseases as they declared 2001–2010 as "Decade of African Traditional Medicine" at the African Summit of Heads of States.

2. Development of policy, legal and regulatory framework: governments should formulate national policies, legal frameworks, and registration according to the guidelines that WHO has provided for the assessment of herbal medicine to establish regional regulatory mechanisms regulating herbal medicine through national expert committees.

3. Promoting scientific research on traditional medicine and collaboration work: scientific research should be conducted on safety, efficacy, and quality of traditional medicine primarily used for the management of priority disease like malaria, HIV/AIDS, sickle-cell anemia, diabetes, and hypertension as proposed by WHO. Collaboration of traditional medicine practitioners with the scientific community can be achieved through staff exchange and training, sharing expensive equipment and joint publications by making partnership arrangements also with the private sector for the integration of traditional medicine.

4. Ensuring that intellectual property rights are protected: intellectual property rights of the indigenous knowledge about traditional medicine should be protected by particular legislation.

5. Disseminating appropriate information to the general public on the use of traditional medicine: appropriate information should be given to the general public to empower them with knowledge and skills for the proper use of traditional medicine through organization of seminars to raise awareness.

6. Providing a good economic environment: the government should ensure a good economic, political, and regulatory environment for local production by traditional herbal practitioners as well as develop industries that can produce standardized remedies to increase access and provide funding for their smooth operations.

There are many factors hindering the development of phytomedicine in Africa:

- Development of drug from its natural source is more difficult than synthetic drug development; formulation of phytomedicine particularly in crude-drug form requires a specialized expert area of training and experience.

- Lack of standardization and quality control of the herbal drugs used in clinical trials.

- The risk of side effects due to:

    - toxicity,

    - over-dosage,

    - interaction with conventional drugs,

    - manufacturing problems such as misidentification of plants,

    - lack of standardization,

    - failure of good manufacturing practice,

    - contamination as a result of field microbial contamination,

    - poor packaging,

    - the bad environmental condition (temperature, light exposure),

    - substitution and adulteration of plants, and

    - incorrect preparation and dosage.

- Imprecise diagnosis and dosage for phytomedicine.

- The lack of communication and collaborative research among orthodox medical practitioners and scientists threaten to lose the ethnomedical knowledge concerning the plant and other aspects of the medicinal system.

- Inadequate randomizations in most studies. Patients are not properly selected and the numbers of patients used in most trials are insufficient for the attachment of statistical significance.

- Problem of serious attention, resource mobilization, commitment, and the required political will.

- There is a wide variation in the duration of treatment using herbal medicine.

- Domestication: it is difficult to convince members of the community to trust phytomedicine after a long use of orthodox medicine.

- There is absence or inadequate record of what is available and many species are becoming extinct because they are not cultivated and protected from indiscriminate harvesting. Also, the traditional healers are of advancing age and dying.

- Unfavorable legislation such as Witchcraft Act.

The quality and stability of phytomedicine is achieved by the use of fresh plants, regulated physical factors like temperature, light, water availability, and cultivation of plants in place of wild-harvested plants, because they show smaller variation in their constituents. The standardization of phytomedicine can also be achieved by the use of chromatography, infrared, and ultraviolet (UV) spectrometry.

The African pharmacognosists, pharmacologists, pharmacists, and physicians have to learn, acquire, document, and use traditional medicine to help curtail the extinction of plants and human resources. Collaborative work can be achieved through staff exchange and training and funding for capital building:

the government should help in funding researches on phytomedicine;

the private sector as well as nongovernmental agencies should help finance researches;

organization of seminars to raise awareness to the general public on the benefits of medicinal plants and remove the perception that scientists are out to harness their knowledge for money making;

abandoning outdated legislation (such as witchcraft Act); and

passing new legislation to protect indigenous traditional knowledge to integrate into the health scheme.

As medicinal plants are going global with increasing demand in the phytotherapeutic market, the following factors must be emphasized in Africa for the development of phytomedicine:

- Emphasis on well-controlled and randomized clinical trials to prove the safety and efficacy of herbal medicine. With the growth of the botanical market, the quality, efficacy, and safety of phytomedicine used in clinical trials have to be improved to produce standardized drugs and to develop novel therapeutic methods with researches on traditional medicines.

- An improvement in the processes of regulation and global harmonization of phytomedicine. The integration of African traditional medicine into the health system should bring harmony between traditional and modern systems of health care with minimum threat to each other.

- Greater emphasis should be placed on collaboration work in order to bring traditional healers closer to scientists by engaging and training them in laboratory work, as well as get information on traditional prescriptions for specific diseases.

- Emphasis has to be placed on domestication, production, biotechnological studies, and genetic improvement of medicinal plants. The domestication of plants will reduce the effects associated with wild-harvested plants, avoid misidentification and field contamination.

Increase the quality of raw materials and yield through genetic breeding and selection.

Production of phytomedicine with resistance to microorganism-induced diseases.

- Detailed legislation on the ownership of intellectual property right has to be made.

Research has shown that a number of traditional medicines are effective therapeutic regimens in the management of a wide spectrum of diseases especially those which may not be effectively managed using Western medicines. Furthermore, inadequate accessibility to modern medicines and drugs to treat and manage diseases in middle- and low-income countries, especially in Africa, may have contributed to the widespread use of TM especially in poor households.

Besides accessibility to traditional healers, TM provides an avenue through which cultural heritages are preserved and respected. Indeed, TM practice is in line with the socio-cultural and environmental conditions of the people who use it in Africa.

Traditional medicine is becoming increasingly popular across the world. However, its growth potential has been understudied and poorly appreciated due to existing global political economy of health and any surrounding informal processes.

In developed countries, on the other hand, factors responsible for the widespread use of TM are beyond accessibility, affordability, and cultural compatibility. According to the World Health Organization anxiety about the adverse effects of chemical drugs, improved access to health information, changing values, and reduced tolerance of paternalism are some of the factors responsible for the growing demand for CAM in developed countries.

Traditional medicine in Africa is contrasted with biomedicine. Most traditional medical theories have a social and religious character and emphasize prevention and holistic features. Traditional medical practices are usually characterized by the healer's personal involvement, by secrecy and a reward system. Biomedical theory and practice show an almost opposite picture: asocial and irreligious with professional detachment. Local communities do not expect that basic health care will improve when traditional healers become integrated into the service. They ask instead for improvement of basic health care itself: more services with better access, more dedication and respect from doctors and nurses, and more medicines and personnel. Fieldwork needs to be done at the community level to arrive at a better understanding and assessment of the community's opinion concerning a possible role of traditional medicine in basic health care.

WHO has been working with African nations to integrate scientific and medical models of health to enhance the potentials of traditional medicine in the control of endemic diseases.

Intelligent application of traditional therapies have made useful contributions to alleviate sickness and suffering in Africa. Efforts should be made to protect plants from going extinct. The people and Orthodox practitioners need to be given appropriate information on phytomedicine. The integration or harmonization of phytomedicine should be developed in such a way to work hand-in-hand with orthodox medicine with minimum threat to each other [24–26].

In the book one, chapter is dedicated to the traditional medicine in Africa.

5. Biofield therapies

Biofield, a term coined during the US National Institutes of Health Conference in 1992, is defined as "a massless interacting field of energy and information that surrounds and permeates living systems."

Biofield therapies are noninvasive therapies in which the practitioner manipulates individuals' energy field in order to stimulate his/her healing responses.

The National Institutes of Health Center for Complementary and Alternative Medicine (NCCAM) has classified energy medicine therapies into two basic categories: bioelectromagnetic-based and biofield therapies. Bioelectromagnetic-based therapies involve the use or manipulation of electromagnetic fields (EMFs).

Biofield modalities, which sense and modulate surrounding "subtle energy fields" and interpenetrate the human body, have existed for thousands of years in a wide range of cultures. The vital energy concepts, which include the Indian term prana, the Chinese term ch'i, and the Japanese term qi, refer to subtle or nonphysical energies. Similar concepts in the West are reflected in the concepts of Holy spirit, or spirit, and can be dated back to the writings in the Old Testament.

A common thread is the development of specific techniques that use subtle energy to stimulate one's own internal (intrapersonal) healing process, by movement-oriented practices such as yoga, tai-chi, or internal qigong, as part of the experience of meditation or prayer.

External (interpersonal) practices specifically use subtle energies for the process of healing another person, including local or proximal practices such as external Qigong, pranic healing, where a healer transmits or guides energy to a recipient who is physically present as well as distance practices where a healer sends energy to a recipient in a different physical location, such as intercessory prayer or distance healing.

A major distinction in biofield therapies involves whether the practitioner engages the patient's biofield with (hands-on) or without physical contact (hands-off). Some modalities such as Reiki and Brennan healing contain techniques that are both hands-on and hands-off (but in close proximity), others such as Johrei and external qigong are practiced with hands at a slightly further distance from the body.

Biofield therapies are inexpensive compared with the costs of other types of therapy and effective to relieve daily life stress of patients by reduction in tension, anxiety, and pain with minimum side effects.

Biofield activities stimulate specific brain areas allowing for differentiation of certain moods. Humans distinguish between pleasant or unpleasant stimuli, based upon biofield information transmitted to the autonomic nervous system, immune system, and the endocrine system, so biofield therapy exert an influence throughout the entire human body.

The majority of practices intended to affect the body's energy flow do not involve devices. Energy Medicine Practices that benefit energy flow and overall energy in the body include certain forms of exercise, mechanical manipulation, pressure, light, sound, scent, touch, position, the use of electrical current or magnetic pulses, or movement to stimulate the body's own energy systems. Exercises of the energy medicine require a trained and authorized instructor to teach the technique (Pilates, Tai Chi) and rely either on manipulation (Alexander Technique, Cranial-sacral), on movement (Feldenkrais, FlexAware), on positions (yoga) or on scents (aroma therapy). Some practices that focus on particular areas of the body require touching (massage, reflexology, Reiki) and some others do not require touching (medical Qigong).

Although science has provided considerable information about how the body works, it cannot explain yet what differentiates living and nonliving matter and why the placebo effect is stronger than most drugs and the roles of attitude and intention, as well as the mechanisms by which they work, which may or may not be related to the placebo effect [27].

Biofield physiology is proposed as a descriptor for the electromagnetic, biophotonic, and other types of spatially distributed fields that living systems generate and respond as integral aspects of the self-regulation and organization of cell, tissue, and whole organism. Individual molecules can act as transmitting and receiving antennae in the mediation of efficient intermolecular communication via single photons.

As a means of information transfer, biophotons have the advantages of extremely high speed and the ability to penetrate through cell membranes that present barriers to the diffusion of molecular signals.

Electrical and magnetic fields, as well as biophotons in the full range from ultraviolet to infrared, are detected during normal physiological activity including the resonance signaling, and the modulation of cell function by specific electromagnetic frequencies.

Biologically generated biofields are a spatially distributed set of forces and physical properties that have the capacity to encode information and exert instructive influences on cells and tissues capable of perceiving and being modified by them [28].

Electric fields generated by the intracellular network of microtubules, centrosomes, and chromosomes play fundamental roles in regulating the dynamics of mitosis, meiosis, and a variety of other cellular activities [29].

Electric charge in motion, whether along a wire or a nerve axon, produces a magnetic field in the surrounding space, a type of biofield. Magnetic fields emanating from the body, although extremely weak relative to the geomagnetic field of the earth, are readily detected by superconducting quantum interference device (SQUID)–based magnetometers [30].

The strongest rhythmic electrical and magnetic fields in the body are produced by synchronous activity of arrays of the heart muscle cells and can be detected, as electrocardiogram (ECG) or magnetocardiogram (MCG). The heart's magnetic field also carries information that can also be detected by other persons or animals and can be recorded up to several feet from the body surface.

An example of the informational potential (bioeffectiveness) of the heart biofields is cardiac-induced entrainment, or frequency locking, detected when the R-waves of one subject's ECG become precisely synchronized with the onset of EEG alpha waves of another subject at a distance of up to 5 feet.

Heart fields may also encode psychoemotional information, as indicated by the 75% accuracy rate in detecting discrete emotional states from patterns of heart rate variability.

The electrical and magnetic fields generated by the composite activity of thousands of brain cells are detected as an electroencephalogram (EEG) or magnetoencephalogram (MEG). At a functional level, the electromagnetic activity of neural assemblies modulates neuron synchronization and circadian rhythmicity and the computational and cognitive processes of the brain. More specifically, weak sinusoidal electric fields enhance and entrain physiological neocortical network activity. Transmembrane currents in neurons also produce local electric fields that induce "ephaptic coupling" (nonsynaptic electrical coupling) between adjacent axons, which influences the synchronization and timing of action potential firing in neurons [31].

Another biofield phenomenon is the coherent, ultraweak photon emissions (UPE), detected from cell cultures and from the body surface. The fluctuations in UPE correlate with cerebral blood flow, cerebral energy metabolism, and EEG activity. Photonic stimulation at one end of a nerve increases UPE at the other end. Nonconventional means of UPE-mediated biosignaling include wave propagation within longitudinally oriented neuronal microtubules and passage through membrane-spanning regions of proteins that may serve as "light pipes" [32].

Numerous nonneural electrical fields have been detected and analyzed, including those arising from patterns of resting membrane potentials that guide development and regeneration, and from slowly varying transepithelial direct current fields that initiate cellular responses against tissue damage. In addition to the high-speed electrical signals conducted along nerve axons, a second communication network, based in ubiquitous epithelial cells, conducts information through varying direct currents, which spread across considerable distances and play key roles in recognizing damage and guiding cell migration necessary for wound healing especially in skin, heart, and cornea as well as in regulating the migration of neuronal path-finding.

Electrical fields—created by either mechanical stress (piezoelectricity) or streaming potentials—in bone, tendons, and fascia regulate the functioning of osteocytes and fibroblasts to adjust the density of supporting tissues in response to loads [33].

Unspecialized "loose" connective tissue, referred as fascia, forms a continuous head-to-toe network surrounding and permeating all tissues and organs. As an extracellular matrix, structured mainly by collagen fibers, fascia provides a supportive and regulatory framework for all organs of the body as it coordinates cellular perception and interpretation of mechanical forces. This extracellular system reaches into the interior of cells via transmembrane bridging molecules known as integrins, which allow information from the fascia to modify cell metabolism and genetic activity. Since collagen structures both conduct and modify photon pulses emitted from biological sources, signaling along collagen fibers serves as a surveillance system of endogenous biofield emission to complement the immune and nervous systems in monitoring tissue health.

Further speculation based on the water-protein relationship along collagen fibers invokes quantum coherence, a state that can occur when all water molecules in a particular domain or region are spinning synchronously, emitting spin or torsion waves. Such spin coherence and quantum coherence enable the collagen matrix to be ultrasensitive to electromagnetic fields in a manner that can be frequency selective due to a quantum phenomenon known as the Larmor Precession [34].

Global coherence is the multilevel integration of diverse biological activities across time and scale accounts for the most salient properties such as long-range order and coordination, rapid and efficient energy transfer, and extreme sensitivity to specific signals.

The receptor system for endogenous and exogenous biofields is a body-wide network that exhibits three types of potential receptor sites: molecular, charge flux, and endogenous field.

An important series of studies on cultured cells identified two examples of the first type of receptor sites—deoxyribonucleic acid (DNA) and the cell membrane—at which exogenous electromagnetic signals exert specific biological effects.

Charge flux sites, the second type of receptor as exemplified by the perturbation of transmembrane calcium fluxes, have been proposed as a generic mechanism by which weak electromagnetic fields affect biological systems.

Low-frequency electromagnetic fields also interact with DNA by accelerating the movement of electrons within the helical arrays of base pairs. While ion channels and ion pumps have major roles in establishing the resting potential of an individual cell, the gap junctions, which are the specialized electrical connections between adjacent cells, allow voltage and current-mediated signals to be propagated across groups of cells. In this manner, spatiotemporal patterns of resting potentials arise to provide bioelectrical guidance during tissue development, regeneration, and cancer suppression [35].

Deletion of the electromagnetic response elements (EMRE) eliminates the ability of the applied electromagnetic field to regulate the target genes, while other genes can be converted from electromagnetic nonresponders to responders by inserting the EMRE at upstream regions.

Sufficient evidence has accrued to consider biofield physiology as a scientific discipline, based on nonlocal, integrated, information-conveying phenomena as well as on emerging molecular details of localized biophysical interactions. Endogenously generated pulses of ultraweak photons, electromagnetic fields, and patterns of distributed membrane voltage are varied forms of physiological activity designated as biofields, each with established properties and proposed biological functions.

While bioelectromagnetics define the mechanisms of local interactions, biofield physiology is more about understanding the integrated, longer-range functions within the whole organism: the former more reductive and the latter more integrative [36].

While the nervous, endocrine, immune, and cardiovascular systems are in continuous intercommunication via electrical and molecular signals, endogenous biofields act as carriers of information between these systems. An example is heart-brain interaction, where several types of cardiac initiated signals appear to exert sequential effects on brain activity. Electromagnetic signals from the heart reach the brain in a relatively instantaneous manner, followed first by a range of neural signals arriving in millisecond timeframes and subsequently by pressure waves and hormonal signals arriving with delays of seconds [37].

Evidence of DNA response elements that respond to specific electromagnetic frequencies, analogous to DNA regions responsive to specific hormones, is an important finding.

Different types of signals mediating rapid/short-acting vs. slower/longer-lasting responses, neurally released adrenaline and hormonally released corticosteroids, respectively, coordinate the stress response. Physiological requirements for ultra-rapid responses may be met by biofields. Raman and infrared spectroscopic techniques are now enabling rapid and sensitive chemical characterization of samples based strictly on the vibrational signatures of the molecules present in a sampling volume. When applied to biological systems, the techniques provide highly complex spectra that document changes taking place in the entire genome, proteome, and metabolome; real-time *in vivo* applications are possible.

Biomarkers, defined as physiological variables that have significant clinical relevance to the population being studied, may include measures of immune, endocrine, psychophysiological, autonomic nervous system (including skin conductance), and other neural functions (including electroencephalography, positron emission tomography). Biomarkers may indicate which physiological systems are affected by biofield therapy but do not necessarily shed light on the pathways by which these changes occur nor on the transduction events by which practitioner activity is converted to patient responses that initiate the cascade of physiological changes [38–42].

An increasing number of physicians and other health care providers have begun integrating biofield therapies into patient care, and a growing number of hospital-based programs offer these modalities to patients. The line between what is "alternative," "complementary," or "integrative" is often blurred when it comes to biofield therapies, their practice, and their use by patients.

In the book, one chapter will be about biofield interpretation.

Healing is a multidimensional process that is strengthened by reducing stress and accessing psychospiritual resources, congruent with their values, beliefs, and philosophical perspectives on life and well-being.

There is a need to re-evaluate all the traditional practices on a scientific base to complement and integrate into the conventional evidence-based medicine. In the last chapter, there is a trial to review and present the current situation and future trends in this integration.

Author details

Cengiz Mordeniz
Namik Kemal University, Tekirdag, Turkey

*Address all correspondence to: cengizmorster@gmail.com

References

[1] Oliveira MJR et al. Fitoterapia no Sistema de Saúde Pública (SUS) no Estado de São Paulo, Brasil. Revista Brasileira de Plantas Medicinais. 2006;**8**(2):39-41

[2] Gurib-Fakim A. Traditional roles and future prospects for medicinal plants in health care. Asian Biotechnology and Development Review. 2011;**13**(3):77-83

[3] Brusotti G, Cesari I, Dentamaro A, Caccialanza G, Massolini G. Isolation and characterization of bioactive compounds from plant resources: The role of analysis in the ethnopharmacological approach. Journal of Pharmaceutical and Biomedical Analysis. 2014;**87**:218-228

[4] Patwardhan B, Joshi K, Chopra A. Classification of human population based on HLA gene polymorphism and the concept of Prakriti in Ayurveda. Journal of Alternative and Complementary Medicine. 2005;**11**:349-353

[5] AyuSoft—a decision support system for Ayurveda. C-DAC (Center for Development of Advance Computing) Pune. Project Sponsored by Ministry of Communication and Information Technology, Government of India. 2004. Available from: www.edac.in

[6] VijayaKumar D, Raghavan KV. Novel Chromatographic Fingerprinting Method for Standardization of Single Medicines and Formulations. Hyderabad: Indian Institute of Chemical Technology; 2002. WO 0246739-EP2 0000991 991-263397CSIR G01N30-88

[7] Patwardhan B, Warude D, Pushpangadan P, Bhatt N. Ayurveda and traditional Chinese medicine: A comparative overview. Evidence-based Complementary and Alternative Medicine. 2005;**2**(4):465-473

[8] Hallman-Mikolaczak A. Ebers Papyrus. The Book of Medical Knowledge of the 16th Century B.C. Egyptians. Archiwum historii i filozofii medycyny/Polskii Towarzystwo Historii Medycyny i Farmacji. 2004;**67**(1):5-14

[9] Msomi NZ, Simelane MBC. In: Builders PF, editor. Herbal Medicine. Rijeka: IntechOpen; 2018. DOI: 10.5772/intechopen.72816. Available from: https://www.intechopen.com/books/herbal-medicine/herbal-medicine

[10] Farhadi F, Khameneh B, Iranshahi M, Iranshahy M. Antibacterial activity of flavonoids and their structure–activity relationship: An update review. Phytotherapy Research. 2019;**33**(1):13-40

[11] Efferth T, Zacchino S, Milen I, et al. Nobel prize for artemisinin brings phytotherapy into the spotlight. Phytomedicine. 2015;**22**(13):A1-A3

[12] Cardini F, Wade C, Regalia AL, Gui S, Li W, Raschetti R, et al. Clinical research in traditional medicine: Priorities and methods. Complementary Therapies in Medicine. 2006;**14**:282-287

[13] WHO. Traditional Medicine. WHO Fact sheet No.134. Revised. December 2008

[14] Colalto C. What phytotherapy needs: Evidence-based guidelines for better clinical practice. Phytotherapy Research. 2018;**32**(3):413-425

[15] Yoshimura M, Enomoto T, Dake Y, Okuno Y, Ikeda H, Cheng L, et al. An evaluation of the clinical efficacy of tomato extract for perennial allergic rhinitis. Allergology International. 2007;**56**:225-230

[16] Newman DJ, Cragg GM. Natural products as sources of new drugs

from 1981 to 2014. Journal of Natural Products. 2016;**793**:629-661

[17] David B, Wolfender JL, Dias DA. The pharmaceutical industry and naturalproducts: Historical status and new trends. Phytochemistry Reviews. 2015;**14**(2):299-315

[18] Dias DA, Urban S, Roessner U. A historical overview of natural products in drug discovery. Metabolites. 2012;**2**(2):303-336

[19] Sheng-Ji P. Ethnobotanical approaches of traditional medicine studies: Some experiences from Asia. Pharmaceutical Biology. 2001;**39**:74-79

[20] Cheng YC. Why and how to globalize traditional Chinese medicine. Journal of Traditional and Complementary Medicine. 2011;**1**(1):1-4

[21] World Health Organization. WHO Traditional Medicine Strategy. 2002-2005. Geneva, Switzerland: World Health Organization; 2002

[22] Bent S. Herbal medicine in the United States: Review of efficacy, safety, and regulation. Journal of General Internal Medicine. 2008;**23**(6):854-859

[23] WHO Global Atlas of Traditional Medicine: Proceedings of an International Meeting; 17-19 June 2003; Kobe, Japan. Kobe, Japan: WHO; 2004

[24] Abdullahi AA. Trends and challenges of traditional medicine in Africa. African Journal of Traditional, Complementary, and Alternative Medicines. 2011;**8**(S):115-123

[25] Okigbo RN, Mmeka EC. An appraisal of phytomedicine in Afrıca. KMITL Science and Technology Journal. 2006;**6**(2)

[26] Elujoba AA, Odeleye OM, Ogunyemi CM. Traditional medical development for medical and dental primary health care delivery system in Africa. African Journal of Traditional, Complementary and Alternative medicines. 2005;**2**(1):46-61

[27] Jain S, Hammerschlag R, Mills P, Cohen L, Krieger R, Vieten C, et al. Clinical studies of biofield therapies: Summary, methodological challenges, and recommendations. Global Advances in Health and Medicine. 2015

[28] Hammerschlag R, Levin M, McCraty R, Bat N, Ives J, Lutgendorf SK, et al. Biofield physiology: A framework for an emerging discipline. Global Advances in Health and Medicine. 2015;**4**(suppl):35-41. DOI: 10.7453/gahmj.2015.015.suppl

[29] Zhao Y, Zhan Q. Electric fields generated by synchronized oscillations of microtubules, centrosomes and chromosomes regulate the dynamics of mitosis and meiosis. Theoretical Biology and Medical Modelling. 2012;**2**(9):26

[30] Steinhoff U, Schnabel A, Burghoff M, et al. Spatial distribution of cardiac magnetic vector fields acquired from 3120 SQUID positions. Neurology & Clinical Neurophysiology. 2004;**59**

[31] McCraty R, Atkinson M, Tomasino D, Bradley R. The coherent heart, heart-brain interactions, psychophysiological coherence, and the emergence of system-wide order. Integral Review. 2009;**5**(2):10-115

[32] Van Wijk R, van der Greef J, van Wijk E. Human ultraweak photon emission and the yin yang concept of Chinese medicine. Journal of Acupuncture and Meridian Studies. 2010;**3**(4):221-231

[33] Ahn A, Grodzinsky A. Relevance of collagen piezoelectricity to "Wolff's law": A critical review. Medical Engineering & Physics. 2009;**31**(7):733-741

[34] Oschman JL. Functional role of quantum coherence in interfacial water. Paper Presented at: Water Conference 2013; Sofia, Bulgaria, October 22-25, 2013

[35] Yang M, Brackenbury W. Membrane potential and cancer progression. Frontiers in Physiology. 2013;**17**(4):185

[36] Rein G. Bioinformation within the biofield: Beyond bioelectromagnetics. Journal of Alternative and Complementary Medicine. 2004;**10**(1):59-68

[37] Uchıda S et al. Effect of biofield therapy in the human brain. The Journal of Alternatıve and Complementary Medıcıne. 2012;**18**(9):875-879

[38] Kosovich J. The regulation of energy medicine. Physics Procedia. 2012;**38**:242-252

[39] Ho MW. The Rainbow and the Worm: The Physics of Organisms. 3rd ed. Singapore: World Scientific Publishing; 2008

[40] Zhao Y, Zhan Q. Electric fields generated by synchronized oscillations of microtubules, centrosomes and chromosomes regulate the dynamics of mitosis and meiosis. Theoretical Biology & Medical Modelling. 2012;**9**:26

[41] Wang E, Zhao M. Regulation of tissue repair and regeneration by electric fields. Chinese Journal of Traumatology. 2010;**13**(1):55-61

[42] Guo A, Song B, Reid B, et al. Effects of physiological electric fields on migration of human dermal fibroblasts. The Journal of Investigative Dermatology. 2010;**130**(9):2320-2327

Chapter 2

Integration of Traditional and Complementary Medicine into Evidence-Based Clinical Practice

Cengiz Mordeniz

Abstract

Traditional and complementary medicine regains popularity not only in developing countries but also in developed countries. Modern medicine often fails to cure and just tries to alleviate the symptoms. The patient feels better as long as the effect of the drug continues but his/her symptoms reappear after the elimination of the drug. In this way, instead of healing the patients, we turn them into life-time drug dependent. Traditional and complementary medicine, being turned scientifically into evidence-based medicine, will change the medical philosophy and treatment such as individualized and holistic approach. Complementary interventions are used together with conventional treatments, whereas alternative interventions are used instead of conventional medicine.

Keywords: complementary medicine, traditional medicine, alternative medicine, traditional Chinese medicine, homeopathy, ayurveda

1. Introduction

Efforts of human beings to find solutions to the diseases, magic and witchcraft and religious beliefs created the basis of traditional medicine. Currently used drugs are mostly sourced from traditional medical products such aspirin, which has been developed as a painkiller from quinine and cocaine has been prescribed in modern medicine for a long time.

Traditional medicine is living harmonized with cultural components where any disease information is passed through the generations. Not only in developing countries, but also in developed countries, traditional practitioners or physicians using traditional methods continue to give primary health services.

2. WHO (World Health Organization) and traditional and complementary medicine

World Health Organization defined traditional medicine as the skills and practices based on health practices, approaches, knowledge and beliefs from various cultures, whether explicable or not, incorporating plant, animal- and mineral-derived medicines, spiritual therapies, manual techniques and exercises, applied singularly or in combination in the maintenance of health and well-being to prevent, diagnose, improve or treat physical and mental illness [1].

The therapeutic outcome is very much related with expectancy, which is created mostly by belief and attitude and results as a placebo, or "meaning response," effect both in modern and traditional medicine.

3. Use of traditional and complementary medicine

From our experience, the most common reasons for using traditional medicine are:

- Economic considerations (it is more affordable).
- The healers know better the sociocultural background of the patients and they are closer to the patient's ideology.
- More concern about the adverse effects of chemical (synthetic) medicines.
- Desire for more personalized health care.
- Greater public access to health information.
- The distances to be covered in some countries.
- The strength of traditional beliefs.
- The shortage of health professionals, particularly in rural areas.
- Low incidence of harmful incidents.
- Perception of traditional and complementary medicine as natural and safe.
- Traditional and complementary medicine is considered natural, safe and non-toxic.

4. Difficulties for the use of traditional and complementary medicine

From our experience, in the use of traditional and complementary medicine, there are problems such as:

- the difficulty in setting the limits
- little knowledge about drug-herbal interactions
- the indeterminant ethical point of view
- the secrecy of healing methods
- the absence of written records
- incorrect diagnosis
- improper dosage
- low hygiene standards

- the lack of a complete code of conduct
- missing scientific and correct instructions, documentation and data base
- insufficient control in labeling and sales
- non-formal structure or organization to train tradi-practitioners
- abuse of traditional medicinal products
- weak labeling
- inefficient control in the market for consumption
- no national nor international convention nor treaty about traditional and complementary medicine

5. Different traditional medicine systems

There are many different systems of traditional medicine, because the philosophy and practices of each are influenced by the prevailing conditions, environment and geographic area [2].

Traditional Chinese medicine (TCM), which is still used in present day health care conserving the holistic approach, is rooted to the ancient philosophy of Taoism and accumulates ancient knowledge dating back more than 2500 years. In this understanding, the human body is considered as a miniature version of the surrounding universe.

TCM encompasses many different practices:

- moxibustion (burning an herb above the skin to apply heat to acupuncture points),
- Chinese herbal medicine,
- tui na (Chinese therapeutic massage),
- dietary therapy and
- acupuncture.

Tai chi: Gentle, dance-like body specific movements or postures, coordinated by breathing, relaxation and mental focus. Tai chi practice improves balance and stability in patients with Parkinson's disease; reduces pain from knee osteoarthritis and fibromyalgia; and increases the quality of life and mood in patients with heart failure.

Harmony between two opposite forces, *yin* and *yang*, maintain the health or imbalance results as a disease. Five elements symbolize all phenomena, and explain the changes in the body functions during any disease. TCM practitioners try to control the yin and yang levels through 12 meridians, to revitalize the (*Qi*), the vital energy, that flows through the body to maintain health [3].

The modalities of CAM vary from culture to culture, and from country to country.

Traditional systems of medicine that exist in other East and South Asian countries, are mostly influenced by TCM and each one has developed distinctive features of its own. For example, Kampo, the system of traditional herbal medicine in Japan, is even covered by the national health insurance plan and is practiced by many licensed medical doctors.

Ayurveda is practiced in India at the national level within the Federal Health System [4].

In Africa up to 90% and in India 70% of the population use traditional medicine for their health care needs [5].

In the United States, under the Dietary Supplement Health and Education Act (DSHEA) of 1994, any herb, botanical and natural concentrate, metabolite and constituent of extract, is classified as a dietary supplement and no additional toxicity studies are required, if the herb has been on the market prior to 1994 [6].

EU has standardized the information and guidelines related to herbal medicines by producing necessary materials, such as monographs on herbs and preparations, guidelines for the collection of materials of herbal origin, the standardization of applications, the identification and quantitative determination of herbal preparations and their complex compositions [7].

More and more people use the traditional medicines even in developed countries because of the positive perception of patients for the use of herbal medicines and satisfaction with therapeutic outcomes coupled with the disappointment with conventional allopathic or orthodox medicine in terms of effectiveness and/or safety [8].

Since herbal extracts may be contaminated or adulterated, their quality control has a direct impact on their safety and efficacy [9].

To isolate each active ingredient from any herb would be time-consuming at a high cost [10].

6. Medicinal plants as the source for pharmaceutical industry

Plants, herbs and ethnobotanicals are the oldest known health care products and vary depending on the ethnological, medical and historical background of each country. Plants and natural sources form the basis of today's modern medicine and pharmacological industry commercially manufacturing drug preparations for health promotion and treatment of disease. About 25% of the drugs prescribed worldwide are derived from plants.

In July 1996, WHO gathered 100 scientific experts from different countries and they made a list of 28 medicinal plants originally prepared by the WHO Collaborating Centre for Traditional Medicine in Chicago, Illinois, United States of America with 28 monographs.

Over the past 100 years, the development and mass production of chemically synthesized drugs have revolutionized health care. For example, three of the top-selling botanical products, namely *Ginkgo biloba, Allium sativum* (garlic) and *Panax ginseng*, originated from TCM.

Currently, herbs are applied for the treatment of chronic and acute conditions and various ailments and problems such as cardiovascular disease, prostate problems, depression and inflammation. In Africa, for example, a traditional herbal medicine, the Africa flower, has been used for decades to treat wasting symptoms associated with HIV [5].

About 960 plant species are used by the Indian herbal industry [11].

Overall international trade in medicinal plants and their products was US$ 60 billion in 2010 and is expected to reach US$ 5 trillion by 2050.

The pharmaceutical industry has come to consider traditional medicine as a source for identification of bio-active agents that can be used in the preparation of synthetic medicine. Some of the medicinal plants have been already phyto-chemically screened and characterized and their antimicrobial, anti-inflammatory, anti-oxidant, anticancer and anti-diabetic activities are being tested by biochemists, toxicologists and pharmacologists, in animal models (in vivo), cell lines, pathogen (bacteria, fungi and viruses), parasites (malaria Plasmodium, sleeping sickness (Trypanosome)) and many others. Such clinical research trials have resulted in safety, efficacy and good pharmacokinetics and pharmacodynamics parameters for potential drugs [12].

About 200 years ago, the first pharmacologically active pure compound, morphine, was produced from opium extracted from seeds pods of the poppy *Papaver somniferum* and then the discovery of penicillin [13].

The expanding herbal product market threatens biodiversity due to the overharvesting of plants. Bad collection and cultivation can lead to the extinction of certain plant species and the destruction of natural resources. It is expected that 15,000 of 50,000–70,000 medicinal plant species are threatened with extinction. The herb-herb and herb-drug interactions challenge the identification of the active compounds in the plants and require research-based evidence, increased awareness and study, as polypharmacy and polyherbacy.

The "traditional" ways of identification and preparation of herbs need to be replaced with more accurate and reproducible methods to ensure the quality, safety and consistency of the product. Given the market value, potential toxicity and increasing consumer demand, particularly in the sick and elderly members, regulation of production and marketing of herbal supplements and medicines bring two main areas of concern such as the international diversity and national policies regarding the regulation of the production and use of herbs (and other complementary medicines) and their quality, safety and scientific evidence in relation to health claims.

7. Regulation of traditional medicine products

In many countries, the herbal medicine market is not adequately regulated, herbal medicines and related products are introduced into the market without any mandatory safety or toxicological evaluation and the products are therefore not registered and controlled by regulatory bodies. There is no effective regulation for manufacturing practices and quality standards. The establishment of regulation and registration procedures is still a problem in both developed and developing countries.

Lack of any documentation system for information, conventions, treaties and decree on medicinal plants led to abuse the uses and sale of products without any instruction or any scientific studies on the active principles and safety and poor labeling.

Some herbal products on the market may be of low quality and suspecting efficacy, but sold without prescription and the potential hazards of such an inferior product may not be recognized. Although herbs may also have undesirable side effects, no set "doses" and herb-drug or herb-herb interactions are possible. There is a belief that herbs, as natural products, are inherently safe without side effects and their efficacy can be obtained over a wide range of doses. The general perception is that herbal remedies or drugs are very safe and devoid of adverse effects. But, herbs have been shown to produce undesirable or adverse reactions causing serious injuries, life-threatening conditions and even death.

In many countries, CAM practice is provided outside the national health care systems and practiced by non-regulated personnel. Therefore, it is usually not monitored by the safety mechanisms and reporting systems of the main-stream regulatory and legislative frameworks.

Generally, health professionals are not trained about the use and the effect of herbal medicines [14].

Analysis of adverse events related to the use of herbal products is more complex than the conventional pharmaceuticals. The evaluation of safety is complicated by factors such as the geographical origin of plant material, different processing techniques, route of administration and compatibility with other medicines. Furthermore, there is a lack of knowledge on taxonomic botany and documentation by most manufacturers of herbal medicines during identification and collection of medicinal plants used for herbal remedies.

8. Definition of medicinal plants

There are different ways to define medicinal plants or herbs. As a result, different approaches have been adopted with regard to licensing, dispensing, manufacturing and trading in order to ensure the safety, quality and efficacy of medicinal plant preparations.

By definition, a dietary supplement is a product ingested to supplement the diet. The dietary ingredients in these products may include vitamins, minerals, herbs or other botanicals.

To eliminate the confusion of common names, the mostly used binomial names (including their binomial synonyms) should be accepted for medicinal plants. For example, *Artemisia absinthium* L., which contains an active narcotic derivative, causing CNS disorders and generalized mental deterioration, has at least 11 different common names. Seven of the common names bear no resemblance to its botanical name.

Heliotropium europaeum (heliotrope), which contains potent hepatotoxic pyrrolidine alkaloids, is often confused with *Valeriana officinalis* (garden heliotrope) known to contain valepotriates with sedative and muscle relaxant properties. Therefore, effective monitoring of safety of herbal medicine requires effective collaboration between botanists, phytochemists and pharmacologists [15].

9. Conclusion

In this book, we have planned to bring together a scientific approach to the development of traditional and complementary medicine, the current use especially in Africa where one of the oldest traditional medical use still exists. The theoretical explanation, which combines with energy and information medicine, explains the bio-field approach as an individualized and holistic therapeutic approach. The interrelation with the modern pharmaceutical industry and the conflicts with the modern medicine are emphasized by the oxidative theory which is quite popular in modern medicine to explain the deviation from homeostasis.

Author details

Cengiz Mordeniz
Tekirdag Namik Kemal University, Turkey

*Address all correspondence to: cengizmorster@gmail.com

References

[1] Bulletin of the World Health Organization. Vol. 86, No. 1. January 2008. pp. 1-80

[2] Rishton GM. Natural products as a robust source of new drugs and drug leads: Past successes and present day issues. The American Journal of Cardiology. 2008;**101**:43D-49D

[3] Xutian S, Zhang J, Louise W. New exploration and understanding of traditional Chinese medicine. The American Journal of Chinese Medicine. 2009;**37**:411-426

[4] Sahoo N, Manchikanti P, Dey S. Herbal drugs: Standards and regulation. Fitoterapia. 2010;**81**(6):462-471

[5] Tilburt JC, Kaptchuk TJ. Herbal medicine research and global health: An ethical analysis. Bulletin of the World Health Organization. 2008;**86**(8):594-599

[6] Harrison RA, Holt D, Pattison DJ, Elton PJ. Who and how many people are taking herbal supplements? A survey of 21,923 adults. International Journal for Vitamin and Nutrition Research. 2004;**74**:183-186

[7] Calapai G. European legislation on herbal medicines: A look into the future. Drug Safety. 2008;**31**:428-431

[8] Cohen PA, Ernst E. Safety of herbal supplements: A guide for cardiologists. Cardiovascular Therapeutics. 2010;**28**:246-253

[9] Ribnicky DM, Poulev A, Schmidt B, Cefalu WT, Raskin I. The science of botanical supplements for human health: A view from the NIH botanical research centers: Evaluation of botanicals for improving human health. The American Journal of Clinical Nutrition. 2008;**87**:472S-475S

[10] Richter RK. Herbal Medicine: Chaos in the Marketplace. New York: Haworth Herbal Press; 2003

[11] Khan S, Chattopadhyay MT, Sarkar D, Chatterjee MC, Sharma T, Dey R, et al. Recent advancements for the evaluation of anti-viral activities of natural products. New Biotechnology. 2009;**25**(5):347-368

[12] Garrido G, Rodeiro I, Hernández I, García G, Pérez G, Merino N, et al. In vivo acute toxicological studies of an antioxidant extract from *Mangifera indica* L. (Vimang). Drug and Chemical Toxicology. 2009;**32**(1):53-58

[13] Li JW, Vederas JC. Drug discovery and natural products: End of an era or an endless frontier? Science. 2009;**325**:161-165

[14] World Health Organization (WHO). National Policy on Traditional Medicine and Regulation of Herbal Medicines. Geneva: Report of WHO global survey; 2005

[15] Zhou Y, Taylor B, Smith TJ, Liu ZP, Clench M, Davies NW, et al. A novel compound fromcelery seed with a bactericidal effect against *Helicobacter pylori*. The Journal of Pharmacy and Pharmacology. 2009;**61**(8):1067-1077

Chapter 3

African Traditional Medicine: South African Perspective

Mmamosheledi E. Mothibe and Mncengeli Sibanda

Abstract

African traditional medicine (ATM) has been used by African populations for the treatment of diseases long before the advent of orthodox medicine and continues to carry a part of the burden of health for the majority of the population. South Africa, as a member state of the World Health Organisation, has been set on the path of institutionalising African traditional medicine. This chapter outlines the processes and progress pertaining to the acceptance and acknowledgement of the role of ATM in health care. It sets out to describe the strides made with regard to the traditional health practitioners' Act and other laws, research in ATM, education of both health care and traditional health practitioners, including the role of collaboration. An overview of the practice of African traditional medicine is provided.

Keywords: African traditional medicine, research, indigenous knowledge

1. Background

The World Health Organisation (WHO) defines traditional medicine (TM) as 'the sum total of knowledge, skills and practices based on the theories, beliefs and experiences indigenous to different cultures, whether explicable or not, that are used to maintain health, as well as to prevent, diagnose, improve or treat physical and mental illnesses' [1]. Therefore, African traditional medicine (ATM) would be the TM indigenous to the different African cultures. Traditional medicine has been used by humankind for the treatment of various diseases since long before the advent of orthodox medicine, and to this day, serve the health care needs of the majority of the world population. According to the WHO, trends in the use of TM and complementary medicines have been increasing [2]. Complementary medicine or alternative medicine (CAM) refers to a broad set of health care practices that are not part of a country's own tradition or conventional medicine and are not fully integrated into the dominant health care system of that country. The terms are used interchangeably with TM in some countries. Other terms sometimes used to describe these health care practices include 'natural medicine', 'non-conventional medicine' and 'holistic medicine' [3]. Traditional medicines (TMs) include herbal medicines, which may be herbs, herbal materials, herbal preparations and finished herbal products [4].

In South Africa (SA), as in countries where the dominant health care system is based on allopathic medicine or where TM has not been incorporated into the national health care system, all other TM not indigenous to SA are termed 'complementary', 'alternative' or 'non-conventional' medicine [5].

African traditional medicine (ATM) is said to be one of the oldest and most diverse of all medicine systems, even though the medicine systems are poorly recorded [6]. African traditional healing is interwoven with cultural practices and religious beliefs and is therefore regarded as being holistic, involving both the body and the mind [6–8]. The WHO has been at the forefront of TM matters, collating information about TM practices worldwide and providing guidelines in advancing the recognition, acceptance and integration of TM into the health systems of member countries.

2. The role of WHO in traditional medicine

The role of the WHO in the recognition, promotion and acceptance of TM dates as far back as 1976; when the World Health Assembly (WHA), which is the governing body of WHO, drew attention to the reserve constituted by those practising traditional medicine. The organisation urged its member states to utilise their traditional systems of medicine. In 1978, at the International Conference on Primary Health Care at Alma Ata, the WHA recommended that governments prioritise the incorporation of traditional health practitioners (THPs) and birth attendants into the health care team and proven traditional remedies into national drug policies and regulation [9].

In 2000, the WHO Regional Committee for Africa adopted a strategy for the African countries, the aim of which was to contribute to the achievement of health for all in the African Region by optimising the use of TM. The strategy urged member states to develop national policies and legislation on traditional medicine and to improve regional and sub-regional collaboration. It also encouraged the member states to take steps to promote and protect traditional medicine nationally. Consequently, the WHO Regional Office for Africa in collaboration with the Department of Essential Drugs and Medicines Policy organised a series of workshops on the regulation of traditional medicines to assist member states to establish mechanisms for evaluating traditional medicines for registration purposes [10].

The WHO African Regional Strategy on TM was adopted in 2001 by the Organisation of African Unity (OAU), now called the African Union. The period 2001–2010 was declared as the Decade for African Traditional Medicine. An annual African Traditional Medicine Day was declared as the 31st of August and is celebrated every year. All stakeholders were requested to prepare a Plan of Action for implementation to meet the objectives of the strategy [11].

The WHO Traditional Medicines Strategy 2002–2005 provided a framework for action to promote the use of TM and Complementary Alternative medicine (TM/CAM) in reducing mortality and morbidity in impoverished nations. The strategy had four objectives, which were

- to integrate TM/CAM into national health care systems, where appropriate, by developing and implementing national TM/CAM policies and programmes,
- to promote the safety, efficacy and quality of TM/CAM by expanding the knowledge base of these remedies and by providing guidance on regulatory and quality assurance standards,
- to increase the availability and affordability of TM/CAM where appropriate, focusing on poorer populations and
- to promote therapeutically sound use of appropriate TM/CAM by providers and consumers [11].

In 2003, the first WHO regional workshop was held in SA where a set of guidelines for the registration of traditional medicines was developed. The minimum regulatory requirements for the registration of TMs were determined [12]. In the same year, resolution WHA56.31 on TM was adopted at the 56th World Health Assembly (WHA). It urged Member States to establish or expand and reinforce existing national drug safety monitoring systems to monitor herbal medicines and other traditional practices [13].

In 2004, 'Guidelines for registration of traditional medicines in the WHO Africa region' were issued. The overall objective of the guidelines was to facilitate the registration, marketing and distribution of traditional medicines of consistent quality in the African Region [12]. The 'Guidelines on the safety monitoring of herbal medicines in pharmacovigilance system' were issued, with the objectives which included promoting the safe and proper use of herbal medicines [4].

By 2004, it had become important for the WHO to obtain detailed qualitative and quantitative information about the prevalence of TM/CAM utilisation. A set of indicators had been proposed by a special consultative meeting of experts in 2001 to measure these aspects. The core indicators were expected to provide answers for who uses TM/CAM, what is used, why is it used and what it costs. The indicators were classified into three main groups. Background indicators included amongst others, aspects such as the total number of prescribers and the total number of TM/CAM providers within and outside of the conventional medicine system. Structural indicators addressed issues such as whether there was official national policy and legislation on TM/CAM. Process indicators provided answers for estimated prevalence of national TM/CAM and estimated prevalence of the five most popular therapies used [14].

The first WHO Congress on Traditional medicine was held in Beijing, China in 2008, to further assess the role of TM/CAM, to review the progress of member states as well as to help these countries integrate TM into their health systems. Part of the Beijing Declaration adopted by Congress stated, amongst others, that

- governments had a responsibility for the health of their people and should formulate national policies, regulations and standards to ensure safe, appropriate and effective use of TM and

- TM should be further developed based on research and innovation.

It also encouraged improved education, clinical inquiry into traditional medicine and improved communication between health care providers [1]. The 62nd WHA held in 2009 endorsed the resolutions of the Beijing declaration and urged member states to adopt and implement the declaration in accordance with national capacities, priorities, relevant legislation and circumstances [15].

The progress report on the Decade of African traditional medicine indicated that many countries popularised TM. They established and strengthened their institutional capacity and developed national policies and regulatory frameworks for TM. There was progress in establishing national programmes and setting up expert committees for the development of TM in their health ministries. By 2010, 22 countries including SA were using WHO guidelines in conducting research on TM for diabetes, hypertension, HIV/AIDs, malaria and sickle cell anaemia. SA and 11 other countries issued market authorisations for TM, ranging from one in Cameroon to more than 1000 in Ghana and Nigeria. Six countries including SA had tools for the protection of Intellectual Property Rights and the TM knowledge [16]. A total of five countries had included TM in their National Essential Drugs list, which was an increase from only one in 1999/2000 to four in 2010.

The WHO Traditional Medicine Strategy for 2014–2023, was released in 2012. It was intended to support member states in, amongst others,

- harnessing the potential contribution of TM to health, wellness and people-centred care,
- promoting the safe and effective use of TM by regulating, researching and integrating TM products, practitioners and practice into health systems where appropriate and
- education and training of TM and CAM practitioners.

This strategy also provided a review of the progress made since the strategy 2002–2005 and it sought to build upon that strategy [5].

The WHO report of 2013 acknowledged that there was growing interest in the world about the use of traditional and complementary medicine (T&CM). The report stated that there had been significant progress in implementing, regulating and managing T&CM in most regions. Member states that had established or developed national policies for T&CM had increased from 39 in 2003 to 69 in 2012 and those who were regulating herbal medicines had increased from 82 in 2003 to 119. Member states had developed regulations on the quality, quantity, accreditation and education structures for T&CM practitioners and conventional medicine practitioners who used T&CM. Member states that provided education on T&CM up to university doctoral degrees had increased from none to 39. Some institutions in the African region, including SA, had included TM in the curricula of health professions students. Some countries across the world had set up national research institutes in the field of T&CM, and those had increased from 56 in 2003 to 73 in 2012 [5].

3. Progress emanating from WHO strategies: legislation and regulation of ATM

Under apartheid and colonial South Africa, the practise of TM was deemed unscientific and illegal. It was considered to be uncivilised, suspect, scientifically unfounded, backward and superstitious [17]. The Witchcraft Suppression Act of 1957 and the Witchcraft Suppression Amendment Act of 1970 declared TM unconstitutional and prohibited practitioners of TM from doing their business [18]. Cooperation between conventional health practitioners (CHPs) and THPs was outlawed by the Medical Association of South Africa in 1953. The prohibition of TM was somewhat based on the conviction that the concept of disease and illness in Africa was generally rooted in witchcraft [19].

An attempt to regulate the practise of THPs was made in 1982 through the promulgation of the Associated Health Service Professions Act of 1982, as amended [20]. This Act set up a registration and licencing scheme for herbalists, chiropractors, homoeopaths, osteopaths and naturopaths, but prohibited their use of the title 'Medical Practitioner'. The province of KwaZulu-Natal was the exception and had a different law on the licencing and control of THPs, which was covered by the KwaZulu Act on the Code of Zulu Law (CZL) of 1981. The CZL allowed for the practise of THPs who were licenced and allowed for them to claim a fee for services rendered [21].

Before the first democratic government in South Africa in 1994, the African National Congress (ANC) submitted in its health plan that THPs would become an integral and recognised part of the health care system in South Africa. It claimed that patients would be granted the right to choose their preferred health care practitioner. At the same time, the ANC realised the need to regulate the practise of THPs in order to protect patients from harmful practices. The ANC health plan further stated the need to promote cooperation and liaison between THPs and allopathic health practitioners [22].

South Africa is a member state of the WHO, the AU and the Southern African Development Community (SADC). All three bodies have accepted resolutions, which urge member states to develop and implement national policies on ATM. The SA government has therefore taken steps towards the official recognition, acceptance and institutionalisation of ATM. Institutionalisation means formalisation and official incorporation of TM into the national health system [23].

The National Drug Policy (NDP) for South Africa of 1996 [24] is amongst one of the first documents to recognise the potential role and benefits of traditional medicine for the national health system. It was aimed at investigating 'the use of effective and safe traditional medicines at primary level' and specified the following, with regard to traditional medicine:

- Traditional medicine would be investigated for its 'efficacy, safety and quality with a view to incorporate their use in the health care system'.

- Marketed traditional medicine will be registered and controlled.

- A National Reference Centre for African Traditional Medicine for African Traditional Medicines (the NRCATM) would be established [25]. The NRCATM, a virtual reference centre, was established in 2003 by the Medicines Control Council (MCC) of the National Department of Health (NDoH) in collaboration with the Council for Science and Industrial Research (CSIR) and the Medical Research Council (MRC). The purpose of the centre was to gather, harness and synthesise information to promote, regulate and register ATMs derived from plants. Some of its functions included the development of a national database of indigenous plants that have been screened for efficacy and toxicity and to test the effectiveness and safety of traditional medicines, so as to protect the public from unproven claims within the traditional medicines sector [25].

A presidential task team was appointed in 2006 to advice on national policy and appropriate regulatory and legal framework regarding TM institutionalisation. This task team drafted the national policy of ATM in SA after consultation with various stakeholders. The Draft National Policy on ATM in SA was gazetted in 2008 [26] and was finally passed as the Traditional Health Practitioners Act (THPA) No. 22 of 2007 [27].

A new directorate, the Directorate: Traditional medicine, (DTM) was established to manage the work related to TM within the NDoH. The vision of the DTM was to advance the contribution of TM to the health and well-being of the population with the mission to facilitate the institutionalisation of TM in the National Healthcare System (NHS) through validation and production of safe and effective TMs and the promotion of TM practice based on a sound legislative and policy framework. The DTM served to coordinate all activities relating to ATM in liaison with units of TM at provincial level and other government departments and statutory research councils such as the CSIR, the MRC and the Agricultural Research Council (ARC) [26].

The THPA (Act 22 of 2007) has now been enacted. This act was purposed to establish an interim Traditional Health Practitioners Council, which was eventually inaugurated in 2013. The council is tasked to provide

- a regulatory framework to ensure the efficacy, safety and quality of traditional health care services and
- the management and control over the registration, training and conduct of practitioners, students and specified categories of the THPs.

The Act makes provisions for a register of all traditional healers and categorises them as sangomas, herbalists, traditional midwives and surgeons. The THPA promotes the training and practice of THPs and proposes the establishment of certified institutions for their training [28]. It further affords THPs the ability to be reimbursed by medical aid schemes for services rendered to patients as well as the authority to issue medical certificates [29].

In accordance with the THPA the regulations for Traditional Health Practitioners were published in the Government Gazette No. 39358, Notice No. 1052 of 3 November 2015 [30]. Although the regulations are in place, there has been a reluctance to take up the registration by THs. One of the challenges is that there is no code of conduct in place; hence, there is no way of determining genuine practitioners from bogus ones for purposes of registration [29, 31]. The non-registration is also as a result of lack of knowledge and understanding of the working of the Act and the registration process. There is also a level of mistrust and suspicion about government's true intention and commitment to ATM and the fear the TM practice being Westernised by conventional health practitioners [17]. Only registered healers have the authority to issue valid sick certificates. Without being registered, the healer cannot provide the service [29]. Although there are still no clear guidelines about this, some employers recognise and accommodate African cultural beliefs and do accept THP-issued sick notes [32].

4. Utilisation and prevalence of traditional medicine in South Africa

The use of TM in developing countries has increased since the 1990s [1]. South Africa is classified as a mixture of both a developing and a developed country [33]. Like other African countries, SA has a pluralistic system of healthcare, in which modern medicine practice coexists with other non-conventional health systems. These include a variety of indigenous systems based on traditional practices and beliefs [34].

African traditional medicine is described in the Draft Policy on African Traditional Medicine for SA as a body of knowledge that has been developed and accumulated over tens of thousands of years, which is associated with the examination, diagnosis, therapy, treatment, prevention of, or promotion and rehabilitation of the physical, mental, spiritual or social well-being of humans and animals [26].

It has been estimated that 80% of Africans use TM, compared to 60% of the world's population in general [1, 7]. The use of ATM by the general public has been reported since years back [35], and ATM is used for many ailments and conditions including for HIV, diabetes mellitus, hypertension, pain, gynaecological disorders, mental disorders and asthma [36–39].

It was reported that in seeking health, people consult traditional healers first before a conventional health provider (CHP) [34, 40–42]. These people may choose to withhold this information when consulting a CHP and not tell that they use ATM [36–38, 40, 43, 44]. A large proportion of the Black population makes use of the dual health care system, in which both the conventional and traditional medicines are demanded depending on the ailment [44]. It was estimated that 72% of the Black African population use ATM, and the average frequency of use per consumer was 4.8 times per year. The number of THPs in SA was estimated to be 350,000 in the early 1990s [45], and in 2009, it was estimated that there is 500 THs for every 100,000 people as opposed to 77 medical doctors for the same population [28].

Recent cases that demonstrated the dualism in health seeking patterns are those involving the therapy of HIV infection. The majority of people living with HIV and AIDS consulted THPs first before visiting conventional medicine practitioners [21, 46]. Many individuals on antiretroviral treatment (ART) also reported the use of TM concomitantly [47–50]. The THs refer their suspected HIV patients to relevant CHPs, yet continue to treat the patients with higher CD4 counts, who are on ARTs [51]. Consumers of ATMs come from all classes and categories of society, including different ages, education levels, religious practices and occupation [38, 44, 52, 53]. The use of TM is a common practice across most sectors of the Black African population and is not merely confined to the poor, rural or uneducated users [44].

There are many reasons cited for the use of TM. Cultural practice is the common reason for use of ATM; others include affordability, availability, accessibility, spiritual and emotional reasons and a general desire for wellness [8, 38, 54]. Affordability means the monetary cost associated with the utilisation of TM treatment, consultation or products thereof. Availability refers to the extent to which TM treatment, provider or products are geographically available to the user [14]. The decision to use a particular medical remedy was dependant on socio-economic variables such as the type of illness; its seriousness, the time it occurs, past experiences of the illness, access to health service, the perceived quality of the service and distrust in clinics [55].

While TM may not always be affordable, it is physically, socially and culturally more available than allopathic treatment. Also, the practice of TM is client centred and personalised, paying due regard to social and spiritual matters that are fundamental to African cultures [56]. In addition to healing of the mind, body and spirit, THPs serve many roles in the community, including counselling, social mediation, cultural education and being custodians of African traditions and customs [41].

5. Research in African traditional medicine

African traditional medicine plays a large role in the management of health, holistically, and in either a preventative, curative and/or palliative nature. Many qualitative studies where THPs are interviewed indicate the diverse range of illnesses and diseases that are managed, from those presenting with physical symptoms to those manifesting psychosocially. There are also claims of cures of conditions without tangible supporting evidence of efficacy, at times, which is one of the reasons why government seeks to protect the interests of members of the public who use the services of traditional health practitioners.

Research on ATM is driven by the desire for new drug discovery as well as the need to validate the use of the ATMs for the specific claims made. There is a dire need to bring the ATM into the scientific understanding of the conventional medicine and CHPs. Most of the TMs have not been well researched scientifically. For CHPs and the scientific community to accept them, there has to be clear understanding of the pharmacology and toxicology of the medicines, which can be explained by the mechanisms of action, the biochemical or physiological pathways the medicines affect and the possible toxicological effects they have. The WHO encourages further research in TM to ascertain the efficacy and safety, as inappropriate use can have negative or harmful effects [13]. A long history of use and simply being natural does not necessarily imply safety. The use of ATMs may cause adverse effects, increase the risk of harmful drug-herb interactions and delay access to effective conventional medicines [13, 57].

Figure 1.
Commonly used and mentioned medicinal plants used as African traditional medicines.

Most of the research in TM is directed at medicinal plants used. The general approach in medicinal plant research is initiated by ethnobotanical studies, followed by experimental laboratory-based processes which include solvent extraction of medicinal plant material with solvents of different polarities, qualitative and quantitative phytochemical screening, bioassay-guided fractionation, isolation of active and/or inactive compounds and structural elucidation of the compounds. The extracts and compounds of interest are then investigated further using various tests for pharmacological and toxicological activities, usually guided by the ethnobotanical use of the plant [58].

The SA government provides a platform across various sectors for research into ATM. The government has formed collaboration with institutions including the Centre for Scientific and Industrial Research (CSIR), Agricultural Research Council (ARC), the Medicines Research Council (MRC), Department of Agriculture (DoA) as well as Department of Science and Technology (DST). Traditional medicine research is conducted at several institutions, facilitated by funding from government through the National Research Foundation (NRF), or other parastatal organisations as part of indigenous knowledge systems (IKSs). Each institution conducts research mainly independently, with a few national and international collaborations present. Hence, the research into ATM is largely uncoordinated. Few flagship programmes are presently running in key institutions with a focus on medicinal plants for the development of immunomodulators, microbicides, anti-diabetic medicines, anti-tuberculosis medicines, antimalarial agents and anti-cancer medicines. Across other institutions, there is a focus on the research in ATM for antihypertensive, anti-diabetic, antimalarial, antimicrobial and anti-HIV medicines.

Over the past few years, ethnobotanical studies in SA have revealed several plants for different conditions. The list (**Figure 1**) includes medicinal plants that have been listed and been investigated for biological, pharmacological and other activities. These plants are used as ATM by communities for conditions, mainly diabetes mellitus, hypertension, HIV infection and others. They also form part of the list of potent medicinal plants and those with potential to be developed into commercial products [59–64].

The toxicology studies of these medicinal plant extracts have been performed. Some of the tests performed include those for anti-inflammatory, antioxidant, antimicrobial, cytotoxic, mutagenic and genotoxic effects [65, 66]. These tests were generally performed after the extraction and concentration of what the researchers believed to be the active compounds within the herbal medicines. Most studies have investigated the toxicity of individual herbs or plant extracts but not mixtures. These are in contrast to how the herbal medicines would normally be used or handled in the traditional medicine practice. However, as recommended by WHO general guidelines for methodologies on research and evaluation of TM, the safety data obtained from *in vitro* tests might not be absolute markers of safety but should be seen as indicators of potential toxicity [67].

6. Collaboration between African traditional medicine practice and conventional research

There is evidence for the need to have collaboration between CHPs and THPs in managing health care. In acknowledging the dualism that exists in seeking health, one of the objectives of the strategic health plan of the NDoH is to form links with THPs and complementary healers.

However, notwithstanding the knowledge that users of ATM do not disclose to their CHPs, there is still a reluctance within the conventional health practice to collaborate with THPs. Several factors hinder collaboration between health practitioners as reported in a few studies. In the biomedical approach, all authority and responsibility are given to the medical practitioner, whereas traditional healing is all-inclusive, taking into account the whole person, as well as the social environment [68]. THPs have indicated their willingness to learn and refer patients to clinics and hospitals; however, CHPs do not share the sentiment this [41, 69]. In one study, CHPs' views were that the two health systems were not compatible with respect to the science involved and the source of knowledge, and that the quality of health care will be compromised if traditional health practitioners are allowed to work in public health facilities. Lack of knowledge about ATM and exposure to it seems to be a major hindrance to opportunities in collaboration with traditional healers [70].

A respectful attitude of open exchange and information is essential for successful collaboration which will impact the health care service positively. The incorporation of traditional healers in modern health care may improve the health care of many South Africans who consult them first before seeking conventional medicine. The collaboration could also help advance the study and research of the medicinal plants that they use [71]. It is recommended that government should develop policy on collaboration, detailing how it should be structured, implemented and monitored. If there were clear guidelines of the implementation thereof, it would facilitate the process and enhance involvement and interaction of all relevant stakeholders [70].

One of the concerns raised particularly by THPs in collaboration with research is the protection of intellectual property and indigenous knowledge. The WHO traditional medicine strategy 2002–2005 advocated for, amongst others, measures to protect knowledge of traditional medicine and plant resources, as well as the intellectual property rights of traditional practitioners [11]. South Africa and other African countries reviewed their legislation to conform to the Trade Related Aspects of Intellectual Property Rights (TRIPS) and the African Union's legislation on protection of traditional medical knowledge of indigenous people [72]. A national indigenous knowledge system (IKS) policy was adopted in 2004, and a national office of IKS was established in 2006. The policy was developed as an overall framework to guide and coordinate the work of different government departments with regard to indigenous knowledge [73]. The Biodiversity Act 10 of 2004, known as 'the Biodiversity Act', provides for the management and conservation of South Africa's biodiversity. This includes the protection of species and ecosystems that warrant national protection, the sustainable use of indigenous biological resources, and importantly, the fair and equitable sharing of benefits arising from bioprospecting involving indigenous biological resources [74].

There are several research groups/centres that show evidence of success in collaboration. At a clinical level, in the province of KZN, THPs treat patients first if they deem appropriate and refer them to the CHPs in the clinics. The patients report back to the THPS after undergoing medical treatment [75, 76]. A collaboration project was initiated on HIV and AIDS, which included researchers, THPs, traditional leadership and national and provincial government officials. The project involved THPs to spread HIV prevention and treatment messages to their patients and the researchers collected data for monitoring and evaluation purposes. The success of the project was the birth of a referral system in which the THPs refer their patients to the clinics, and clinics being able to refer patients to THPs who have special knowledge in certain diseases. This relationship has resulted in clinics appointing a THP to their board of directors [77].

Collaboration in research is showing great progress in terms of bringing ATM into understanding of science fraternity. The Department of Science and Technology (DST) primarily ensures effective coordination and responsiveness to industry and academia. The DST works closely with communities in flagship projects that have resulted in patenting and commercialisation of novel products. Through its Bio-economy Strategy the department has led to the development, formulation, patenting and commercialisation of products that are nutraceuticals (Moringa iced tea and Moringa vitamin water) and cosmeceuticals (skin tone evener) [78].

The MRC in collaboration with the University of Cape Town (UCT), developed the TM database, TRAMED III which was aimed at gathering research on the TM and making the information available to the public. A practical guidebook for traditional healers in Primary Health Care Principles was compiled and published, in collaboration with THP groups [47, 79]. CSIR is collaborating with THPs to investigate the efficacy and toxicity of the TMs based on how THPs use them. This collaboration has led to the identification of a number of traditional herbal remedies. One success venture is that CSIR has partnered with Afriplex, in research, development, and patenting of a product from *Elephantorrhiza elephantina* for the treatment of male pattern baldness [80]. Both the CSIR and the MRC have benefit-sharing policies that guarantee compensation for THPs and communities for commercialising medicinal plant products [18].

Within its Indigenous Knowledge Systems unit, the University of the Free State is involved in research in medicinal plants used for the treatment of malaria, HIV, cancer and diabetes. The collaboration with THPs has produced indigenous teas, which are at the level of commercialisation [81]. When the university hosted the South African Society for Basic and Clinical Pharmacology Conference in 2017, it was noteworthy to have the participation of staff members from the Traditional Healers' Organisation, who form part of their IKS unit collaboration. The conference therefore marked a new beginning for closer association of researchers and THPs, by allowing engagement on the same platform.

7. Education

There are no formal institutions in SA for the training in TM as a practice. The training of a THP is a formal and painstaking process, which is initiated in a unique manner for each prospective trainee. The training period may be months to one or more years depending on how fast the trainee learns the trade, and on the calling. The calling to become a healer may manifest as a form of illness, which would then be verified by a qualified diviner [82, 83].

The THPA 22 of 2007 stipulates the initiation and setting up for an institution of training. All healer practice training is performed by trainees under the guidance of a practising and experienced healer. The minimum standards of training are clearly outlined, indicating the expected minimum duration of training, competencies, and age for entry [27].

In line with the call for indigenisation/decolonisation/Africanisation of the curricula, there is an acceptance and recognition of the importance of teaching medical or health science students in higher institutions of learning about ATM. This need was critically driven and accelerated by the students' protest action which started in 2017. Part of the call for the movement was that the curriculum in universities was still shaped and influenced and geared towards a Western and conventional education system and model. The training of health sciences and medical students is intended to prepare the students as health professionals who would largely operate within a health system in which the recipients of the services are within a population that practice traditional medicine in addition to the orthodox medicine. Therefore, there is a dire need to prepare the students adequately and optimally for the real situation that they will be released into. Prior to 2017, only one of the eight medical schools in the country had incorporated TCAM in to the curriculum, despite calls by government to introduce concepts of TM in the teaching [84].

Exposure and training at undergraduate level regarding the traditional health system are recommended as it would facilitate collaboration with THPs [70]. ATM is indigenous knowledge in medicine. More institutions have now incorporated ATM and CAM into their curricula. The challenge will occur with content development and delivery modes. Current academics may be inadequately equipped in terms of knowledge on ATM, while there also may be negative perceptions and prejudices. Hence, the teaching on ATM in the early stages will require academics to be objective and creative in providing the information. The student populations may have diverse perceptions, opinions and attitudes about ATM. The knowledge base of the students would also be variable, ranging from those who have no knowledge at all, to those who have experience on ATM. It would require the lecturer to have the confidence and competence of teaching sensitive topics such as ATM.

8. Overview of African traditional medicine practice

According to the THPA, traditional medicine is the sum total of skills and practices based on beliefs and experiences indigenous to African cultures, that are used to prevent, diagnose, improve or treat physical and mental illnesses.

Traditional medicine is an object or substance used in traditional health practice for the diagnosis, treatment or prevention of physical or mental illness; or well-being in human beings; and traditional health practice means the performance of a function, activity, process or service based on traditional philosophy, that includes the utilisation of TM. Traditional healing is associated with herbs, remedies and advice from a traditional healer, with a strong spiritual component. For this reason, it is impossible to separate African traditional healing from African spirituality [47, 83]. African spirituality encompasses belief and worship to God, and reverence and acknowledgement of ancestors. Ancestors are compassionate spirits of the departed blood-relatives of an individual, and may involve a whole lineage spanning generations [83, 85, 86]. They are revered but not worshiped as one would pray to God and serve to mediate between the living and God [87]. They are regarded as custodians of the lives of future generations and therefore occupy a position of dignity and respect within their descendants [86]. Anecdotal evidence has been noted, where a client or a trainee healer inadvertently discovers their true identity in terms of family name and origin, something that would cause consternation in families, particularly where this had been kept secret from the individual. This is equivalent to how DNA tests confirm blood relations in orthodox medicine. African traditional healing or African indigenous healing in SA is closely associated with African indigenous churches, most of which practice Christianity. This is largely as a result of early influence by Western missionaries and colonisation. Nonetheless, African traditional healing embraces God as the Supreme Being and Creator, the main pillar of the universe and is the same God that Christianity and other religious practices believe in [83, 86].

Communication with ancestors is facilitated by a THP who would also guide on how to specifically communicate depending on the purpose and the ritual that may be required. The consultation occurs at different time periods and differ from group to group [83, 86]. The ritual usually involves the slaughtering of an animal, either chicken, goat, sheep or cow, depending on the purpose, the significance or simply the instruction from ancestors. The slaughtering is important as the blood signifies the connection between the individual and the ancestors. It represents the eternal bond between the ancestors and their descendants. For that reason, the slaughtering has to be done properly, according to specific instructions and at the right place which often is the homestead and can never be at an abattoir [83]. This spiritual healing provides a sense of security, anchoring and validates their identity of the descendants and a sense of belonging and purpose in life [86].

Ancestors may extend beyond the individual or personal level, as in when dealing with community issues [83]. In such cases, reverence will be directed at the ancestors of the village, the tribe or the country. In SA, the government has erected a memorial, the Freedom Park in Pretoria in recognition of the role of ancestors to the living. Within the park there are several sacred sites that are assigned the meaning and importance of the park. The Wall of Names is inscribed with the names of those who lost their lives in the various conflicts that took place in South Africa, from pre-colonial wars to the South African War (Anglo Boer War), World War I, World War II and the liberation struggle. The Garden of Remembrance, *Isivivane*, is the spiritual resting place of those who played a role in the freedom and liberation of South Africa. It is a tranquil space that invites reflection and prayer and visitors entering are required to take off their shoes as a sign of respect. Ancestral rituals are occasionally performed in this garden, including when the remains of those who perished in foreign countries during the struggle are repatriated. The Sanctuary has been created to provide meditative space which permits introspection and quiet contemplation. Guests are allowed to light a candle to honour their ancestors and remember the many South Africans who sacrificed their lives to ensure freedom for all [88].

According to the THPA, THPs are classified as diviners, herbalists, traditional surgeons and traditional birth attendants [18, 27, 83]. A traditional healer is 'a person who is recognised by the community where he or she lives as someone competent to provide health care by using plant, animal and mineral substances and other methods based on social, cultural and religious practices' [9]. There are estimated to be between 200,000 and 300,000 traditional healers in South Africa [89]; with the healer-patient ratio of 1:500–1200, as compared to the medical doctor-patient ratio of 1:40000. Faith healers and prophets are no longer classified under the THPA as they are more church-oriented practice and are therefore legislated separately [41]. The diviners use divination to communicate with their ancestral spirits (and their patient's) to diagnose their patients' misfortunes or ailment, which may be done using different physiological, psychiatric and spiritual conditions [83, 85]. Diviners are said to be spirituality experts as they diagnose and define illness, its origin and the reason in terms of African belief systems [85, 90]. The herbalist practices the art of healing by administration of relevant herbs and plant material. The herbalist has extensive knowledge of plants, herbs, insects, animals, birds and snakes [34]. Diviners and herbalists are called *iSangoma* and *iNyanga* in *isiZulu*, respectively, and both are African shaman [87]. The terms are commonly used although other SA indigenous languages have their own names. Traditional surgeons are generally males qualified, accredited, trusted, experienced and recognised to perform circumcision on boys [34, 83]. Traditional birth attendants are older women who have years of experience as midwives, also provide care and advice during pregnancy, and may provide maternal and child care [34].

The trigger for consultation of a traditional healer may be illness, manifesting physically and or psychosocially. Actual experience of symptoms of illness (e.g., headaches, acute or chronic body pain), an injury (trauma) or heightened senses of anxiety and restlessness, unhappiness, depression or uncertainty are some of the factors that may spur consultation. Some of the reasons for consulting a traditional healer include accessibility, affordability, trust in the healer, culture and beliefs, spiritual and emotional reasons, time devoted to patient by the healer, emphasis on cure rather than prevention in conventional treatments, limitations of conventional treatments in dealing with chronic health problems and a general desire for wellness [54].

For diagnosis, a healer may use an appropriate medium, depending on the practice of the diviner. The diviner may silently observe the patient, or use a particular object as a symbol of the connection to the patient such as a lit candle, a glass of water or an item of clothing, etc. The patient can also do self-diagnosis by informing the healer what the patient or others have observed. Common diagnostic methods of the diviner include spirit possession, the use of divination bones and dreams. Spirit possession is when the ancestral spirit occupies the healer, and therefore, the healer serves as a conduit for messages. Divination bones are a set of various small objects ranging from shells, dice, tiny animal bones to plant kernels, which represent various elements in the spiritual context. The bones, when thrown by either the diviner or the patient, fall in a particular pattern intended by the ancestors and specific to the patient. The diviner then interprets this pattern as applicable to the patient or the situation. Diviners are also adept at interpreting dreams as ancestors may send messages in dreams [87]. Treatment in traditional healing may be in various forms. ATMs (products) are derived from various plant parts such as leaves, stem, bark, roots; herbs, animal sources and other materials such as water and ash [83]. Generally, a healer issues the ATM with instructions of use or may direct a patient where to collect the medicines. The practice of using plants and animal products is widespread in Africa and trading in plants and animal parts for ATM is common in many African towns and cities [91]. ATM may be a single item or combination of either plant or animal products or other materials.

Medicinal plants used as ATM by THPs and communities for the treatment of various conditions including diabetes mellitus [92], cancer [93], asthma [39, 94]), tuberculosis [95], for use by HIV infected persons [96] and mental disorders [39] have been compiled and listed through many ethnobotanical studies. As an example, some of the plants used for diabetes mellitus, such as *Aloe ferox*, have also been investigated scientifically for anti-diabetic properties. The leaf, as well as the sap from the leaf, of *Aloe ferox* is commonly used in ATM, and *in vitro* and *in vivo* studies have shown that it has insulin-increasing activity which may be the mechanism by which it reverses hyperglycaemia [59, 62, 97, 98]. *Ubulawu*, a traditional medicine prepared from roots of *Silene bellidioides* and stem of *Helinus integrifolius*, is used to cleanse the body and the mind. This medicine is therefore said to provide both physical and psycho-spiritual healing. THPs use it to open their intuition and enhance their dreaming, to increase their learning ability and bring forth their gifts of healing; while ordinary people use it for dreaming and to increase general health and energy [85].

A diverse range of animal products is used as ATM [99] as evidenced by the notable trade in TM markets in SA [100], and other countries such as Mozambique that provide cross border trading thereof [101]. The animal species traded include various species of birds, reptiles, mammals and amphibians. This trading has raised concern as some of the animals are listed for conservation, and the trading is extensive but not well understood and not controlled [100]. Animal parts used may include one or combination of parts such as the skin, tail, meat, hair, bones, teeth, fat, glands and faecal pellets. The conditions managed include physical and mental illnesses as well as antenatal care [91]. The highest reported sales in the ATM market in Faraday, Johannesburg were of crocodiles, pythons and baboons [100]. In traditional healing, crocodile fat is mixed with ground bark of *Cryptocarya latifolia* for the treatment of chest pain, and with ashes of burnt powdered bark of *Euclea natalensis* for treatment of abnormal growths [102].

ATM may be administered through various routes depending on ailment. The oral route includes drinking (concoction, decoction, infusion), eating (powdered plant material mixed with food and chewing (parts of plant—leaf, bark, root and spitting out the chaff). Decoctions, which are concentrated extracts, are prepared by adding cold water to the amount of plant material required and allowing it to boil. Infusions are prepared like tea, by steeping the herbs in boiling water and taking it either warm or cold. Enemas and emesis (purging) are ways of detoxing to get rid of toxic substances that are unsuitable to stay long in the gastrointestinal tract. *Ubulawo*, mentioned earlier, is used as an emetic and purgative [85, 86]. Inhalational route involves smoking dried, burnt plant material or steaming (vapour). Topical administration includes bathing in water in which the ATM is suspended or applying a paste on the skin, in the same way that a lotion would be applied. A poultice is a paste from the crushed fresh plant material, usually mixed in water and then applied on the required part of the body [6]. Subcutaneous administration involves making tiny incisions in the skin and applying the medicine as a paste by rubbing it into the cut [87]. This ensures entry of the medicine into the system. Other modalities of administration of treatment include prayer to God and/or devotion to ancestors, drumming, chants and dances; and reassurance and rest [82, 86]. The ATM may also be worn as a charm, amulet or bracelet [34].

9. Commercialisation of African traditional medicine

ATM was generally associated with herbs, remedies and advice from diviners or healers with strong spiritual and cultural components [8]. Usually, ATM would be obtained from traditional healers or by self-collection on advice from a healer or someone knowledgeable about the medicine. Commercialisation of indigenous medicinal plants is a process that has been growing gradually since the early 1990s, with the aim of developing the known medicinal plants into various health products [90]. Commercialisation is the process of introducing a new product into the commerce, i.e., making the product available on the market. Due to commercialisation, some ATMs are available ready for use from various retail outlets. These include grocery stores, *muthi* markets, health shops, street side vendors, supermarkets and over-the-counter (OTC) in pharmacies. *Muthi* is the Zulu word for medicine. The producers of commercially available TM are the retail *muthi* shops, health shops that specialise in herbal medicines, pharmaceutical manufacturers and Laissez faire manufacturers [44, 52]. Some medicinal plants used for common ailments have been developed into medicinal products and commercialised by large manufacturing pharmaceutical companies such as Phyto Nova Pty (Ltd) and Aspen Pharmacare. These medicines are available as formal processed and standardised preparations in modern packaging and in dosage forms such as capsules, ointments, tablets, teas or tinctures. Some of the medicines are derived from plants that are common household traditional remedies such as *Aloe ferox, Artemisia afra*, *Harpagophytum procumbens*, *Hypoxis hemerocallidea*, *Lippia javanica*, *Sutherlandia frutescens* and *Pelargonium sidoides* [103, 104].

There is a growing informal or semi-formal trade sector, in which a large number of medicinal plants are sold as crude, unprocessed plant material. The traditional medicines trade is huge and growing, generating billions of rands per annum in South Africa in various sectors [44]. There are no clear statistics providing an indication of how many ATM (*muthi*) shops are available currently, but these shops form part of the commercial landscapes of many metropolitan centres of South Africa. With commercialisation, many of the ATMs are being prepared and processed as herbal mixtures or concoctions readily available for sale to the public. Most of them are developed, packaged and branded similar to over-the-counter (OTC) medicines and known herbal supplements [103]. Some of them may be in the form of coloured solids, brightly coloured and scented liquids, capsules and incense sticks [105]. Street vendors, owners of *muthi* shops, THPs and TM users were interviewed and agreed that the use and trade of plants for medicinal use are no longer confined to traditional healers, but had entered both the formal and informal entrepreneurial sectors of the SA economy [106].

These commercial HMs are popularised by advertisements and promotions in both print and electronic media. Social networks, the internet, radio, television and newspapers have provided a new platform for marketing of ATM by THPs [107]. The market for CHMs will continue growing as it facilitates the use of ATMs through self-medication. It may be that it is convenient in that both time and money are saved in bypassing the need for consultation of a THP. They are generally sold in retail outlets, muthi shops, by streetside hawkers as well in pharmacies. However, even with modern packaging and marketing practices, the CHMs have not been tested for efficacy and safety, and their labels do not provide adequate information about medicines [105, 108]. An observation made was that there were inconsistencies and deceptive tendencies regarding the labelling and the actual contents; hence, the labels cannot be trusted about the contents of the containers. Nevertheless, the lack or incompleteness of information or about these medicines does not dissuade their use by the public or their sale by the various available retailers/vendors, including pharmacies. It was noted that the market for CHMs may be driven by the desire for urbanised populations to use TM, but having neither the time nor the resources to produce the medicines [105]. The use of the CHMs may therefore be solely based on manufacturer claims and marketing. There might also be the indirect connotation that since the HMs are sold in pharmacies; then, they are guaranteed for safety and efficacy. There may also be the perception that as natural products they signify purity and safety [108]. Nonetheless, of course 'natural' does not necessarily imply safety [23].

The increasing availability of the concoctions or the commercial herbal mixtures used as ATMs have raised the interest of researchers, as they are produced for the public and sold as ready-made mixtures; and their market seems to be growing. Researchers have evaluated some of these for various activities and effects; including toxicology studies. Four concoctions tested displayed antimicrobial activity, variably against *Bacillus subtilis*, *Staphylococcus aureus*, *Escherichia coli* and *Klebsiella pneumoniae*, and *Candida albicans* [109], and had anti-cyclooxygenase activity (anti-inflammatory). Other CHMS were found to have inhibitory and stimulatory effects on isolated and non-isolated human neutrophils and platelets; effects which were also observed variably at low and high concentrations of the herbal mixtures. These effects were observed in neutrophils and platelets from healthy diabetes mellitus, hypertension and asthma patients. The effects may signify beneficial roles in the conditions, yet they also imply possible complications, which may be harmful [110, 111]. A batch of concoctions were confirmed to have different levels of cytotoxicity and mutagenicity [39, 112]. Other concoctions had moderate antibacterial activity against *Enterobacter cloacae*, *Enterobacter aerogenes*, *Escherichia coli*, *Klebsiella pneumonia* and *Citrobacter braakii* [113].

A great concern has been raised with regard to their purity or possible contamination either during production, storage or dispensing, which can be through adulteration by chemicals and biological contamination [114, 115]. Biological contamination refers to impurities in medicinal herbs and their preparations and products and may involve living microbes such as bacteria and their spores, yeasts and moulds, viruses, protozoa, insects (their eggs and larvae) and other organisms [116]. Microbial contamination was found in herbal medicines sold in the Nelson Mandela Metropole (Port Elizabeth) and in Johannesburg, which may pose a major health risk to patients. Thus, there is a need for introduction of guidelines for the quality control in preparation and storage of ATMs and community awareness about the potential health risks [114, 115]. Herbal concoctions bought directly from private traders and streetside sellers were evaluated and found to have been contaminated by *Enterobacter cloacae*, *Enterobacter aerogenes*, *Escherichia coli*, *Leclercia adecarboxylata* and *Citrobacter braakii*, which are motile bacteria and hence are more damaging to cells [113]; and *K. pneumoniae* and *P. fluorescens*, which may negatively affect the health of those using them [117]. The concoctions were also contaminated by commonly occurring yeasts—*Stephanoascus ciferrii* and *Cryptococcus laurentii*. These findings suggest an urgent need for regulation

Freshly collected plant material at a market place

Crude processed plant material ready for sale

Dried, processed plant material in a THP room

Ready mixed concoctions or herbal mixtures ready for sale in a shop

Figure 2.
Plant material in various stages of the production: from collection to point of use as a ready-to-use mixture.

measures of the handling of the herbal mixtures during production, storage and dispensing. The South African Health Products Authority (SAHPRA), through its working group, is working on the draft framework for the regulation of ATMs produced for bulk sale [118], and these studies should serve as a guide for fast-tracking the processes (**Figure 2**).

10. Current African traditional medicine issues: the status of *Cannabis sativa*

In September 2018, the Constitutional Court of SA announced that the prohibition of for private use and consumption of cannabis, as well as cultivation for personal use is against the constitutional rights of citizens to privacy, thereby legalising the use of cannabis privately [119]. The ruling, however, made it clear that public consumption, distribution or selling and use by minors is illegal. This pronouncement has changed the whole landscape of cannabis use, including the use for medicinal purposes. Beforehand, in 2016, cannabis had been legalised for medicinal use, and a team of experts had identified priority areas for research which would advance its use and benefit healthcare [120]. In 2017, the SAHPRA (which was still MCC) had issued guidelines on growing of cannabis for medicinal use and for manufacture of related pharmaceutical products. For cultivation, production, manufacture and use of medicinal cannabis products, a licence has to be applied for and be obtained from SAHPRA, as well as a permit from the Department of Health [121]. Only one institution, the Department of Pharmacology, University of the Free State, has been granted licence to perform systematic research on medicinal cannabis thus far [82].

Cannabis has been reported for use as TM for various conditions including diabetes mellitus [92], cancer [93], asthma [39, 94], tuberculosis [95], for use by

Figure 3.
Cannabis sativa.

HIV-infected persons [96] and other conditions [39]. It is used as a decoction or infusion, alone or mixed with other herbs and taken daily as a preventative medicine, and it would be obtained from THPs [39, 94]. The legalisation certainly provides an environment for continued use of the plant as ATM and obligates that more research should be done to explore further benefits of medicinal cannabis. Hence, the regulations around granting licence for research may need to be reviewed in light of the ruling (**Figure 3**).

11. Conclusion

The use of African traditional medicine by the public will persist and keep growing. ATM plays a role that cannot be completely substituted by conventional medicine; hence, it will remain as a part of the healthcare option available to the population as long as it is accessible. Legislation and policies are in place to facilitate institutionalisation of the TM, although there is slow progress of applications of the various laws. It is pertinent that SAHPRA has to accelerate the framework of regulation of the production of ATMs for bulk sale as the market is growing rapidly, increasing the potential risks to the public. While there is notable progress and benefit in institutional research and collaboration, there is a great need to provide guidelines and regulation for collaboration at primary health care and clinical level. Proper education of conventional health providers about African traditional medicine and the role of traditional health practitioners will facilitate understanding and trust between the two practices and benefit the health care service. The complexities that are delaying the enactment of the Traditional Health Practitioners Act should be addressed, so that the registration, training and acceptance of THPs will be realised.

Conflict of interest

None.

Author details

Mmamosheledi E. Mothibe* and Mncengeli Sibanda
Sefako Makgatho Health Sciences University, Pretoria, South Africa

*Address all correspondence to: mamza.mothibe@smu.ac.za

References

[1] World Health Organization. Traditional Medicines. 2008. Available from: http://www.who.int/mediacentre/factsheets/fs134/en/ [Accessed: January 08, 2011]

[2] World Health Organisation. 2013. A Global Brief on Hypertension: World Health Day. 2013. Available from: http://www.who.int [Accessed: June 10, 2015]

[3] World Health Organisation. Available from: http://www.who.int/medicines/areas/traditional/definitions/en [Accessed: June 15, 2015]

[4] World Health Organisation. Guidelines on Safety Monitoring of Herbal Medicines in Pharmacovigilance Systems. 2004. Available from: http//apps.who.int/medicinedocs/documents/s14215e [Accessed: June 06, 2015]

[5] World Health Organisation. The WHO Traditional Medicine Strategy 2014-2023. 2012. Available from: http://www.who.int/mediacentre/factsheets/fs134/en/ [Accessed: June 15, 2015]

[6] Gurib-Fakim A. Medicinal plants: Traditions of yesterday and drugs of tomorrow. Molecular Aspects of Medicine. 2006;**27**:1-93

[7] Helwig D. African traditional medicines. Alternative Medicine Encyclopedia. 2005 Available from: http://www.answers.com/topic/traditional-african-medicine [Accessed: February 02, 2015]

[8] Truter I. African traditional healers: Cultural and religious beliefs intertwined in a holistic way. South African Pharmaceutical Journal. 2007;**74**(8):56-60

[9] WHO. Declaration of Alma-Ata International Conference on Primary Health Care, Alma-Ata, USSR, 6-12 September 1978. 2000. Available from: www.who.int/hpr/NPH/docs/declaration_almaata.pdf [Accessed: December 16, 2015]

[10] World Health Organisation. Regional Office for Africa, Progress Report on Decade of Traditional Medicine in the African Region. 2011. Available from: www.afro.who.int/ [Accessed: June 08, 2015]

[11] World Health Organisation. The WHO Traditional Medicine Strategy 2002-2005. 2002. Available from: http://www.who.int/mediacentre/factsheets/fs134/en/[Accessed: October 08, 2015]

[12] World Health Organisation, Regional Office for Africa. Regional HFA Policy for the 21 Century in the African Region: Agenda 2020: Harare. 2004. Available from: www.afro.who.int [Accessed: June 15, 2015]

[13] World Health Organization. Traditional Medicine, Highlights of the 56th World Health Assembly. 2003. Available from: www.who.int/features/2003/05b/en [Accessed: December 16, 2015]

[14] World Health Organisation. Legal Status of Traditional and Complimentary/Alternative Medicine: A Worldwide Review. Geneva: WHO; 2001. 16 p

[15] World Health Organization. Sixty-Second World Health Assembly: Traditional Medicines. 2009. Available from: http://www.who.int/mediacentre/factsheets/fs134/en/ [Accessed: June 08, 2015]

[16] Kasilo OMJ, Trapsida JM, Mwikisa CN, Lusamba-Dikassa PS. An overview of the traditional medicine situation in the African region. The African Health Monitor. 2011;**14**:7-15

[17] Le Roux-Kemp AA. Legal perspective on African traditional medicine in South Africa. The Comparative and International Law Journal of Southern Africa. 2010;**43**(3):273-291

[18] Hassim A, Heywood M, Berger J. Health and democracy: A guide to human rights, health law and policy in post-apartheid South Africa. 2007;**20**(2):201-224. Available from: http://www.alp.org.za/publications/healthanddemocracy/Chapter7.pdf [Accessed: November 18, 2015]

[19] Abdullahi AA. Trends and challenges of traditional medicine in Africa. African Journal of Traditional, Complementary and Alternative Medicines. 2011;**8**(S):115-123

[20] Freeman M, Motsei M. Planning healthcare in South Africa—Is there a role for traditional healers? Social Science and Medicine. 1992;**34**:1183-1190. DOI: 10.1016/0277-9536(92)90311-D

[21] Mngqundaniso N, Peltzer K. Patients consulting traditional health practitioners in the context of HIV/AIDS in urban areas in Kwazulu-Natal, South Africa. African Journal of Traditional, Complementary and Alternative Medicines. 2008;**5**:370-379

[22] African National Congress (ANC). A National Health Plan for South Africa. 1994. Available from: https://www.sahistory.org.za/sites/default/files/a_national_health_plan_for_south_africa.pdf. [Accessed September 20, 2015]

[23] World Health Organisation. The World Health Report. Traditional medicine Fact Sheet No 134. 2003. Available from: http://www.who.int/mediacentre/factsheets/fs134/en/ [Accessed: October 08, 2010]

[24] Department of Health. National Drug Policy for South Africa. Pretoria: Government Printer, 1996. p. 26. Available from: http://www.doh.gov.za/docs/policy/drugsjan1996.pdf [Accessed: November 18, 2015]

[25] Department of Health. 2004. Available from: http://www.doh.gov.za/mediaroom/index.html [Accessed: September 20, 2015]

[26] Department of Health. Draft policy on African traditional medicines for South Africa. Government Gazette, 31271 (906). 2008. Available from: http://www.doh.gov.za/docs [Accessed: September 20, 2015]

[27] Republic of South Africa. South African Traditional Health Practitioners Act 22 of 2007. Pretoria: Government Printers; 2007

[28] Cook CT. Sangomas: Problem or solution for South Africa's health care system. Journal of the National Medical Association. 2009;**101**(3):261-265

[29] Tshehla B. Traditional health practitioners and the authority to issue medical certificates. South African Medical Journal. 2015;**105**(4):279-280. DOI: 10.7196/SAMJ.9217

[30] Republic of South Africa. Traditional health practitioners regulations. Government Gazette 39358 (1052). 2015

[31] Fokazi S. Still No Traditional Healers Registered. 2015. Available from: https://www.iol.co.za/news/politics/still-no-traditional-healers-registered-1841714

[32] Mbatha N, Street RA, Ngcobo M, Gqaleni N. Sick certificates issued by South African traditional health practitioners: Current legislation, challenges and the way forward. South African Medical Journal. 2012;**102**(3 Pt 1):129-131

[33] Rautenbach C. Institutionalisation of African traditional medicine in

South Africa: Healing powers of the law. Journal of Contemporary Roman-Dutch Law (THRHR). 2011;**74**(1):28-46

[34] Meissner O. Traditional medicine and its accommodation in the South African national health care system with special attention to possible statutory regulation [LL. D thesis]. Unisa; 2009

[35] Peltzer K. Utilisation and practice of traditional/complementary/alternative medicine (TM/CAM) in South Africa. African Journal of Traditional, Complementary and Alternative Medicines. 2009;**6**(2):175-185

[36] Malangu N. Self-reported use of traditional, complementary and over-the counter medicines by HIV-infected patients on antiretroviral therapy in Pretoria, South Africa. African Journal of Traditional, Complementary and Alternative Medicines. 2007;**4**(3):273-278

[37] Gqaleni N, Moodley I, Kruger H, Ntuli A, McLeod H. Traditional and complementary medicine. South African Health Review [Health Systems Trust]. 2007. Available from: www.hst.org.za/uploads/files/Ch 12 [Accessed: August 15, 2015]

[38] Otang WM, Grierson DS, Ndip Ndip RN. Prevalence and perceived benefits and effectiveness of herbal medicine in the management of symptoms of opportunistic infections in HIV/AIDS patients in the Eastern Cape, South Africa. African Journal of Biotechnology. 2011;**10**(83):9458-19463

[39] Sorsdahl K, Stein DJ, Grimsrud A, Seedat S, Flisher AJ, Williams DR, et al. Traditional healers in the treatment of common mental disorders in South Africa. Journal of Nervous and Mental Disease. 2009;**197**(6):434-441. DOI: 10.1097/NMD.0b013e3181a61dbc

[40] De Villiers FPR, Ledwaba MJP. Traditional healers and paediatric care. South African Medical Journal. 2003;**93**(9):664-665

[41] Moshabela MM. Reasons given by caregivers for administering African herbal medicines to children at St Rita's hospital in Sekhukhune District of Limpopo Province, South Africa [M Med thesis]. University of Limpopo; 2008

[42] Dambisya YM, Tindimwebwa G. Traditional remedies in children around Eastern Cape, South Africa. East African Medical Journal. 2003;**80**(8):402-405

[43] Peltzer K, Phaswana-Mafuya N, Treger L. Use of traditional and complementary health practices in prenatal, delivery and postnatal care in the context of HIV transmission from mother to child (PMTCT) in the Eastern Cape, South Africa. African Journal of Traditional, Complementary and Alternative Medicines. 2009;**6**(2):155-162

[44] Mander M, Ntuli L, Diederichs N, Mavundla K. Economics of the traditional medicine trade in South Africa. In: South African Health Review. Durban: Health Systems Trust; 2007. pp. 189-199. Available from: www.hst.org.za/uploads/files/Ch 13 [Accessed: January 20, 2010]

[45] Bodecker G. WHO. Planning for cost-effective traditional health services. In: Traditional Medicine, Better Science, Policy and Services for Health Development. Awaji Island, Japan: Hyogo Prefecture; 2000. pp. 31-70

[46] Morris K. Treating HIV in South Africa—A tale of two systems. The Lancet. 2001;**357**:1190

[47] Richter M. Traditional medicines and traditional healers in South Africa. Discussion paper prepared for the Treatment Action Campaign and AIDS

Law Project. 2003. Available from: http://www.tac.org.za/Documents/ResearchPapers/Traditional_Medicine_briefing.pdf [Accessed: December 19, 2015]

[48] UNAIDS. Collaborating with Traditional Healers for HIV Prevention and Care in Sub-Saharan Africa: Suggestions for programme managers and field workers. UNAIDS Best Practice Collection. 2006. Available from: http://data.unaids.org/ub/Reort/2006/jc0967 [Accessed: October 20, 2015]

[49] Babb DA, Pemba L, Seatlanyane P, Charalambous S, Churchyard GJ, Grant AD. Use of traditional medicine by HIV-infected individuals in South Africa in the era of antiretroviral therapy. Psychology Health and Medicine. 2007;**12**:314-320. DOI: 10.1080/13548500600621511

[50] Peltzer K, Friend-du Preez N, Ramlagan S, Fomundam H, Anderson J, Chanetsa L. Antiretrovirals and the use of traditional, complementary and alternative medicine by HIV patients in KwaZulu-Natal, South Africa: A longitudinal study. African Journal of Traditional, Complementary and Alternative Medicines. 2011;**8**(4):337-345

[51] Audet CM, Ngobeni S, Wagner RG. Traditional healer treatment of HIV persists in the era of ART: A mixed methods study from rural South Africa. BMC Complementary and Alternative Medicine. 2017;**17**:434. DOI: 10.1186/s12906-017-1934-6

[52] Cocks M, Moller V. Use of indigenous and indigenized medicines to enhance personal wellbeing: A South African case study. Social Science and Medicine. 2002;**54**:387-397

[53] Hughes GD, Aboyade OM, Clark BL, Puoane TR. The prevalence of traditional herbal medicine use among hypertensives living in South African communities. BMC Complementary and Alternative Medicine. 2013;**13**:38

[54] Truter I. Why do people use complementary and alternative medicine? South African Pharmaceutical Journal. 2002;**69**(2):44

[55] Mbelekani NY, Young-Hauser AM, &. Coetzee JK. The Sangoma or the healthcare center? health-seeking practices of women living in the Mangaung Township (Bloemfontein, South Africa). Qualitative Sociology Review. 2017;**13**(1):210-227

[56] UNAIDS. Report on the Global HIV/AIDS Epidemic–June 2000. Available from: http://www.unaids.org/en/resources/documents/2000/20000619_2000_gr [Accessed: September 20, 2015]

[57] Abbott R. Documenting Traditional Medical Knowledge. 2014. Available from: https://www.wipo.int/export/sites/www/tk/en/resources/pdf/medical_tk.pdf [Accessed: October 08, 2017]

[58] Sasidharan S, Saravanan D, Chen Y, Sundram KM, Latha LY. Extraction, isolation and characterization of bioactive compounds from plants' extracts. African Journal of Traditional, Complementary and Alternative Medicines. 2011;**8**(1):1-10

[59] Odeyemi S, Bradley G. Medicinal plants used for the traditional management of diabetes in the Eastern Cape, South Africa. Pharmacology and Toxicology. Molecules. 2018;**23**:2759. DOI: 10.3390/molecules23112759

[60] De Wet H, Ramulondi M, Ngcobo ZN. The use of indigenous medicine for the treatment of hypertension by a rural community in northern Maputaland, South Africa. South African Journal of Botany. 2016;**103**:78-88

[61] Mahomoodally MF. Traditional medicines in Africa: An appraisal of

ten potent African medicinal plants. Evidence-Based Complementary and Alternative Medicine. 2013. DOI: 10.1155/2013/617459 [Accessed: June 15, 2016]

[62] Street RA, Prinsloo G. Commercially important medicinal plants of South Africa: A review. Journal of Chemistry. 2013:1-16. DOI: 10.1155/2013/205048 [Accessed: June 10, 2017]

[63] Omoruyi BE, Bradley G, Afolayan AJ. Ethnomedicinal survey of medicinal plants used for the management of HIV/AIDS infection among local communities of Nkonkobe Municipality, Eastern Cape, South Africa. Journal of Medicinal Plants Research. 2012;**6**(19):3603-360823. DOI: 10.5897/JMPR12.541

[64] Deutschländer MS, Lall N, van de Venter M. Plant species used in the treatment of diabetes by South African traditional healers: An inventory. Pharmaceutical Biology. 2009;**47**(4):348-365

[65] Taylor JLS, Rabe T, LJ MG, Jager AK, Van Staden J. Towards the scientific validation of traditional medicinal plants. Plant Growth Regulation. 2001;**34**:23-37

[66] Fennel CW, Lindsey KL, McGaw LJ, Sparg SG, Stafford GI, Elgorashi EE, et al. Assessing African medicinal plants for efficacy and safety: Pharmacological screening and toxicology. Journal of Ethnopharmacology. 2004;**94**:205-217

[67] World Health Organisation. General Guidelines for Methodologies on Research and Evaluation Traditional Medicines. 2000. Available from: whqlibdoc.who.int/hq/2000/WHO_EDM_TRM_2000.1 [Accessed: October 20, 2016]

[68] Botha CF. The Sangoma and the MD: The clash of Western medical science and traditional medicine in South Africa. Phroniman. 2004;**5**(2):39-50

[69] Campbell-Hall V, Petersen I, Bhana A, Mjadu S, Hosegood V, Flisher AJ. Collaboration between traditional practitioners and primary health care staff in South Africa: Developing a workable partnership for community mental health services. Transcultural Psychiatry. 2010;**47**(4):610-628. DOI: 10.1177/1363461510383459

[70] Nemutandani SM, Hendricks SJ, Mulaudzi MF. Perceptions and experiences of allopathic health practitioners on collaboration with traditional health practitioners in post-apartheid South Africa. Journal of Primary Health Care and Family Medicine. 2016;**8**(2):1007. DOI: 10.4102/phcfm.v8i2.1007

[71] Nethathe GD, Russell SL. Traditional medicine use and the anaesthetist. Southern African Journal of Anaesthesia and Analgesia. 2014;**20**(6):221-225

[72] Chatora R. An overview of the traditional medicine situation in the African Region. African Health Monitor. 2003;**4**(1):4-7

[73] Ngang CC, Ageh PA. Intellectual property protection of African traditional medicine within the legal framework of the right to development. African Journal of International and Comparative Law. 2019;**27**:3

[74] Gericke N. Muthi to medicine. South African Journal of Botany. 2011;**77**:850-856

[75] Health Systems Trust Annual Report. 1998. Available from: http://www.hst.org.za/publications/Annual%20Reports/annualrep98.pdf [Accessed: October 15, 2016]

[76] Tugendhaft A. Medical Pluralism and HIV/AIDS in South Africa: What

are the barriers to collaboration between Traditional Healers and Medical Doctors? Master of Arts (Health Sociology) thesis. Johannesburg: University of the Witwatersrand; 2010. Available from: http://wiredspace.wits.ac.za/bitstream/handle/10539/8470/ [Accessed: October 16, 2018]

[77] Gqaleni N, Hlongwane T, Khondo C, Mbatha M, Mhlongo S, Ngcobo M, et al. Biomedical and traditional healing collaboration on HIV and AIDS in KwaZulu-Natal, South Africa. Universitas Forum. 2011;**2**(2):1-9

[78] Department of Science and technology. 2014. Available from: www.dst.org.za

[79] South African Medical Research Council. 2016. Available from: www.samrc.co.za

[80] Council for Scientific Research. 2018. Available from: www.csir.co.za

[81] UFS Directorate Research Development. UFS Research Report 2016/2017. 2017. Available from: www.ufs.ac.za/docs/librariesprovider41/research-resources-documents/ufs-research-report-2016-2017 [Accessed: October 20, 2018]

[82] Vilakazi T. Ephehlweni: Expert advice. In: Healer. Vol. 2. Johannesburg: Bangoma News; 2010:20-22

[83] Mokgobi MG. Understanding traditional African healing. African Journal of Physical Health Education Recreation and Dance. 2014;**20**(2):24-34

[84] Chitindingu E, George G, Gow J. A review of integration of traditional, complementary and alternative medicine into the curriculum of South African medical schools. BMC Medical Education. 2014;**14**(1):1-5

[85] Sobiecki JF. The intersection of culture and science in South African traditional medicine. Indo-Pacific Journal of Phenomenology. 2014;**14**(1):1-10. DOI: 10.2989/IPJP.2014.14.1.6.1238

[86] Edwards SD. A psychology of indigenous healing in southern Africa. Journal of Psychology in Africa. 2011;**21**(3):335-348

[87] Cumes D. South African indigenous healing: How it works. Explore. 2013;**9**(1):58-65

[88] Available from: http://freedompark.co.za/elements-of-the-park/elements-overview/spirituality.html [Accessed: December 19, 2018]

[89] Van Niekerk J. Traditional healers formalised? South African Medical Journal. 2012;**102**:105-106

[90] Semenya SS, Potgieter MJ. Medicinal plants cultivated in Bapedi traditional healers homegardens, Limpopo Province, South Africa. African Journal of Traditional, Complementary and Alternative Medicine. 2014;**11**(5):126-132

[91] Ntiamoa-Baidu Y. Wildlife and food security in Africa. FAO Conservation Guide 33. Available from: http://www.fao.org/docrep/w7540e/w7540e00.htm [Accessed: December 19, 2018]

[92] Afolayan AJ, Sunmonu TO. *In vivo* studies on antidiabetic plants used in South African herbal medicine. Journal of Clinical Biochemistry and Nutrition. 2010;**47**:98-106

[93] Koduru S, Grierson DS, Afolayan AJ. Ethnobotanical information of medicinal plants used for treatment of cancer in the Eastern Cape Province, South Africa. Current Science. 2007;**92**(7):906-908

[94] Semenya SS, Maroyi A. Plants used by Bapedi traditional healers to treat asthma and related symptoms

in Limpopo Province, South Africa. Evidence-Based Complementary and Alternative Medicine. 2018:33. Article ID 2183705. DOI: 10.1155/2018/2183705

[95] Lawal IO, Grierson DS, Afolayan AJ. Phytotherapeutic information on plants used for the treatment of tuberculosis in Eastern Cape Province, South Africa. Evidence-Based Complementary and Alternative Medicine. 2014:11. Article ID 735423. DOI: 10.1155/2014/735423

[96] Peltzer K, Friend-du Preez N, Ramlagan S, Fomundam H. Use of traditional complementary and alternative medicine for HIV patients in KwaZulu-Natal, SA. BMC Public Health. 2008;**8**:255. DOI: 10.1186/1471-2458-8-255

[97] Balogun FO, Tshabalala NT, Ashafa AOT. Antidiabetic medicinal plants used by the Basotho tribe of Eastern free state: A review. Journal of Diabetes Research. 2016;**2016**:1-13. Article ID 4602820. DOI: 10.1155/2016/4602820

[98] Loots DT, Pieters M, Islam MS, Botes L. Antidiabetic effects of *Aloe ferox* and *Aloe greatheadii* var. davyana leaf gel extracts in a low-dose streptozotocin diabetes rat model. South African Journal of Science. 2011;**107**(7/8):1-6. DOI: 10.4102/sajs

[99] Available from: http://www.zuluculture.co.za/healing/animal-cures [Accessed: December 19, 2018]

[100] Whiting MJ, Williams VL, Hibbitts TJ. Animals traded for traditional medicine at the faraday market in South Africa: Species diversity and conservation implications. Journal of Zoology. 2011;**284**:84-96

[101] Williams VL, Moshoeu TJ, Alexander GJ. Reptiles sold as traditional medicine in Xipamanine and Xiquelene Markets (Maputo, Mozambique). South African Journal of Science. 2016;**112**(7/8):1-9. DOI: 10.17159/sajs.2016/20150416

[102] Grace OM, Prendergast HDV, Jäger AK, Van Staden J. Bark medicines used in traditional healthcare in KwaZulu-Natal, South Africa: An inventory. South African Journal of Botany. 2003;**69**(3):301-363

[103] Van Wyk BE. The potential of South African plants in the development of new medicinal products. South African Journal of Botany. 2011;**77**:812-829

[104] Van Wyk BE. A broad review of commercially important southern African medicinal plants. Journal of Ethnopharmacology. 2008;**119**:342-355

[105] Ndhlala AR, Van Staden J. Smokescreens and mirrors in safety and quality of herbal medicines: A case of commercialized herbal preparations. South African Journal of Botany. 2012;**82**:4-10

[106] Dold AP, Cocks ML. The trade in medicinal plants in the eastern Cape Province, South Africa. South African Journal of Science. 2002;**98**:589-597

[107] Bonora F. The mordenity/traditional interface amongst urban black South Africans: An investigation of the current themes [MSc thesis]. University of South Africa; 2001

[108] Ndhlala AR, Stafford GI, Finnie JF, Van Staden J. Commercial herbal preparations in KwaZulu-Natal, South Africa: The urban face of traditional medicine. South African Journal of Botany. 2011;**77**:830-843

[109] Ndhlala AR, Stafford GI, Finnie JF, Van Staden J. In vitro pharmacological effects of manufactured herbal concoctions

used in KwaZulu-Natal South Africa. Journal of Ethnopharmacology. 2009;**122**(1):117-122

[110] Mothibe ME, Kahler-Venter C, Osuch E. *In vitro* effects of a commercial herbal medicine used as African traditional medicine on human neutrophils. African Journal of Traditional, Complementary and Alternative Medicines. 2017;**14**(3):51-60

[111] Mothibe ME. The *in vitro* investigation of the toxicity of commercially available herbal mixtures used as African traditional medicine in Pretoria, Gauteng, South Africa [PhD thesis]. Pretoria: Sefako Makgatho Health Sciences University; 2017

[112] Ndhlala AR, Anthonissen R, Stafford GI, Finnie JF, Verschaeve L, Van Staden J. *In vitro* cytotoxic and mutagenic evaluation of thirteen commercial herbal mixtures sold in KwaZulu-Natal, South Africa. South African Journal of Botany. 2010;**76**:132-138

[113] Matotoka MM, Masoko P. Phytochemical screening and pharmacological evaluation of herbal concoctions sold at Ga Maja Limpopo Province. South African Journal of Botany. 2018;**117**:1-10

[114] Van Vuuren S, Williams VL, Sooka A, Burger A, Van der Haar L. Microbial contamination of traditional medicinal plants sold at the faraday muthi market, Johannesburg, South Africa. South African Journal of Botany. 2014;**94**:95-100

[115] Govender S, Du Plessis-Stoman D, Downing TG, Van de Venter M. Traditional herbal medicines: Microbial contamination, consumer safety and the need for standards. South African Journal of Science. 2006;**102**:253-255

[116] Kosalec I, Cvek J, Tomić S. Contaminants of medicinal herbs and herbal products. Archives of Industrial Hygiene and Toxicology. 2009;**60**(4):485-501. DOI: 10.2478/10004-1254-60-2009-2005

[117] Matotoka M, Masoko P. Evaluation of herbal concoctions sold at Ga Maja (Limpopo Province) in South Africa and *in vitro* pharmacological evaluation of plants used to manufacture the concoctions. Journal of Evidence-Based Complementary and Alternative Medicine. 2017;**22**(4):805-815. DOI: 10.1177/2156587217727112

[118] SAHPRA Strategic Plan 2018-19 to 2022-23. Available from: https://www.hpasa.co.za/wp-content/uploads/2018/06/SAHPRA-Strategic-Plan-2018-19-to-2022-23.pdf [Accessed: September 15, 2018]

[119] Constitutional Court of South Africa Case CCT 108/17. 2018. Available from: http://www.saflii.org.za/za/cases/ZACC/2018/30.pdf [Accessed: October 10, 2018]

[120] Department of Health. Cultivation of Cannabis and Manufacture of Cannabis-Related Pharmaceutical Products for Medicinal and Research Purposes. 2017. Available from: www.sahpra.org.za/documents/84a71af62.44_Cannabis_growth_Feb2017_v1_for_comment.pdf [Accessed: October 30, 2018]

[121] Augustine TN, Cairns CJ, Chetty S, et al. Priority areas for cannabis and cannabinoid product research in South Africa. African Journal of Primary Health Care and Family Medicine. 2018;**10**(1):a1711. DOI: 10.4102/phcfm.v10i1.1711

Chapter 4

La Gioconda Effect and Self-Organizing Factors of Biofield

Silvio Leite Monteiro da Silva

Abstract

The intriguing trajectory of Mona Lisa painting by Leonardo da Vinci and its attraction to people is similar to the story of Homeopathy: an unprecedented brief description of La Gioconda effect. Homeopathy continues to move through the use, textual production, and new modalities of using its medications. The development of the "Self-Organizing Factors of Biofield" (BioFAO) with complex of seven homeopathic medicines will be presented applied in life sciences. BioFAO corroborates Hering's Healing Laws and obstacles to healing. Regardless of the point of view, innovations happen in Homeopathy, either in the stricto sensu or in the lato sensu of its classical known terminology. The *La Gioconda* effect mysteriously maintains Homeopathy for centuries, as well as Mona Lisa's trajectory of magnetism over people, which made it become the most famous painting of mankind. Quantum theory brings elements that can fundamentally be connected with phenomena applied with living beings already observed.

Keywords: *La Gioconda* effect, Self-Organizing Factors of Biofield, BioFAO, ultradilutions, complex homeopathy, biological phenomena, proving, obstacles to cure, animals

1. Introduction

Homeopathy is a therapeutic tool of traditional knowledge that is used to intervene in health-disease processes of people, animals, and plants. The subject arouses high motivational interest in the society as described in this chapter due to the *La Gioconda* effect. This knowledge area is in constant evolution and is developing new ways to use the dynamized medicines; therefore, it was identified the use of the word Homeopathy in the *stricto* and *lato sensu*. So, it will be addressed the Self-Organizing Factors of Biofield (BioFAO) methodology, the context while an acting agent in the biofield and considerations regarding the traditional knowledge of the Homeopathy. Specifically, it will be described the brief history of its development, mentioned the medicines used in this complex, the main criteria for the choice of the potencies, the effect of the Healing Laws with BioFAO, obstacles to healing, use in animals, and how the discovery of each one of the information presented is related to the *La Gioconda* effect.

1.1 *La Gioconda* effect

The magnetism by interest that Mona Lisa, of Leonardo da Vinci, arouses in the people became this painting one of the most known and famous in the history of contemporary humanity. The passing of time and all the controversial facts that happened put the icon of the Renaissance in a position of world highlight. Located outside Italy, the museum that exhibits it attracts more than 8 million people annually and has this work popularly as the most sought. The great sure is the hearty discussions about several historical and technical aspects and investigations on this painting. However, its riddles continue to be discussed until today, and, mysteriously, continue without a consensual answer to facts that have not been unraveled, even after five centuries of the existence of a painting seemingly simple [1–6].

The particular people's fascination with a painting can be identified in parallel in other exciting subject presented to the world by Samuel Hahnemann in the secular scale: Homeopathy. It is estimated that there are over 200 million people in treatment according to the most recent data; from this total, 100 million would be European citizens, the equivalent to 1/3 of the total of all this population. The curiosity and facts that happened keep warm discussions in the society since its conception for over 200 years and keep it alive by looking for clinical practice, researches, and material published about the Art of Healing—its main lemma. For an example, Homeopathy and Jesus are the only two controversial web pages of Wikipedia in four languages simultaneously. Today, it is considered the second therapeutic system more used in the world, but, maybe, the most discussed between the several areas and professional specialities, the society and scientists, queens and kings [7–14].

Figure 1.
La Gioconda effect: the face with smile of Hahnemann in overlapping to Mona Lisa. The painting of Mona Lisa was so famous that artists made several works based on this painting, to the point of the people desiring to put their face overlapping the original through editing techniques by digital image cropping, and has become so popular that it is considered an internet meme [15].

Given the proportion of the different areas, both the subjects were presented by Renaissance's geniuses. Their achievements crossed centuries, wars, natural disasters, and significant cultural changes of the humanity. When considering the chronological magnitude, both by fragility of a work of art painted on canvas as by culture and habits as abstract as the desire or the need to use homeopathic medicines, surely both would have been forgotten by time if they were so uninteresting to the people. Besides, the several opinions of each one of these work get to be very rich and, at the same time, distinct. What we note is exactly a common highly magnetic effect between the two works. Both arise actions of defenders and disdainful, historical and heroic acts, they cause movements to their audience and to their nonaudience. Inspired in the original name in Italian of the work Mona Lisa, the phenomenon of attraction by the interest of the people in the Homeopathy, as the brief description, is called of "*La Gioconda*" effect (**Figure 1**).

2. Literature review

2.1 The biofield for science

Biofield is notoriously recognized in the science and by USDA. It is assessed in the people and even in animals in the worldwide by noninvasive diagnosis methods, like, for example, since the discovery of X-rays, laureate with the Nobel Prize in Physics of 1901; and more, the discovery of the electrocardiogram, laureate with the Nobel Prize of 1924 in Physiology or Medicine, and, yet, by discovery of computed tomography (CT or CAT scans), which was laureate with the Nobel Prize of 1979 in Physiology or Medicine, by discovery of magnetic resonance imaging, which was laureate with the Nobel Prize of 2003 in Physiology or Medicine; the thermography and, also, electroencephalography. Today, all these analyses continue to be used in the clinical routine with principles essentially identical, only improved by technological developments, such as the increase of the image resolution and programming resources. However, all these analyses are based on detection of consolidated occurrences in the tissues of the living organisms, basically limited to the use as diagnosis techniques. This detection happens from perturbations of the respective field spectra assessed in the living organism in each one of these analysis mentioned [16–21].

The use of electromagnetic waves has a precedent in therapeutic treatments of the predominant Western medicine and in several procedures. Nevertheless, it is used very specifically and without considering the assessment of the traditional knowledge of the biofield, as much as potentially interferes in it. Examples of use of these techniques are the electroconvulsive therapy (formerly called of electroshock therapy) used in the psychiatry area; electroshock as containment or stunning method; execution agent by electrocution (popularly known as eletric chair), and euthanasia by focused beam microwave irradiation. Currently are used neuromuscular blockers in the treatments to avoid the typical muscle contractions that mark the society by disgusting aspect of some of these methodologies, with the voltage adjustment and more refined exposure places. However, these techniques, whatsoever objective and scientific arguments that exist, they present adverse effects or reactions by part of professionals of the area, on other specialties and areas, and by lay population regarding to acceptance, because several of them are highly rejected, contrary to the *La Gioconda* effect [22–24].

Recently, the use of fields started to be applied in an interventionist way with apparent damage containment in relation to the alleged benefits, sometimes in a noninvasive way. Examples of these techniques are radiotherapy for intervention in neoplastic processes; implant of cardiac pacemakers; electrical cardioversion in emergencies of cardiac arrhythmias; shock wave lithotripsy (SWL); laser beams for surgical interventions; cardioversion with blue light (optogenetics); neuromodulation; but they continue disregarding the biofield for treatment as the traditional knowledge can do [25–31].

The technology of dynamization of the pharmacotechnic homeopathic extracts the active principles of a certain material substance that can have mineral, vegetal, or animal origin. The dynamization is a sequential procedure that uses subsequent dilutions and shaking in liquid and dilutions and triturating in solid. According to the constant of Avogadro, theoretically would exist low probability in the molecular presence of the original active principle after the dynamization of 12c or 24d. It is considered that there is a significant detection of molecules originating from the homeopathic matrix only in low dynamizations, but not in high or ultra-high dynamizations. The active principle of a dynamized medicines, which are also called of high dilutions or ultra high dilutions, possibly will be the information impregnated in the base substance that can be in one weak wave or several weak waves, not knowing if it would be, necessarily, exclusively an electromagnetic wave [32–34].

2.2 From classical Homeopathy to Self-Organizing Factors of Biofield

La Gioconda effect provided to Homeopathy its ongoing study and development by doctors and researchers from classical traditional methodologies. The discovery of new medicines increased the descriptions of material medica; introduced new ways to use the medicines known, integrated new practices applied to the criteria of choice of the medicines; sequences of medicines; enlargement of the details of assessment of the healing laws; software for electronic repertorization; assessments in animal models; and development of several protocols based on homeopathic medicines and pharmacology [35–40].

The use of the word Homeopathy has a new dimension beyond the *stricto sensu*, according to the Hahnemannian traditional literature. However, to strengthen the area, notoriously multiprofessional and interdisciplinary, the use of the word Homeopathy in *lato sensu* should be considered when we refer to the therapeutic methods that use dynamized medicines, created from the disciples of Hahnemann. An example of this evolution would be the methodologies of the 12 Schüessler Tissue Salts, use of biotherapeutics, application of the dynamized medicines by anthroposophy, homotoxicology, and the Self-Organizing Factors of Biofield (BioFAO) [10, 11, 32–35, 39].

Several of these therapeutic opinions aggregated traditional knowledge of the many therapeutic styles around the world to the use of homeopathic medicines. One of the main contributions for Homeopathy was preceded, in the history of the science of Western Medicine predominant, the use of a single medicine. The context was a time in which many medicines were used and, without knowing what were specifically their effects, became panaceas and, ultimately, did not know which one was healing or intoxicating. Nowadays, the several opinions of Homeopathy in *lato sensu* consider the use of more than one dynamized medicine or more than one potency, and it is different of the phenomenon of abuse by a polymedication like happens in the allopathy by use of more than five drugs simultaneously. Motivated by context of use of the classic method with a single medicine, known by unicism, emerges the will for a robust and consensual therapeutic platform. So, the methodology currently described as BioFAO was developed during the clinical practice of the Doctor Miria de Amorim and her team of collaborators in Rio de Janeiro (Brazil), coherent with the phenomenon described in the *La Gioconda* effect in the beginning of this chapter [11, 12, 39–41].

Amorim describes her methodology in 2018 (available in Portuguese), the summary of the history, inspirations, and motivations that boosted the development of BioFAO. The first parameter assessed was the biological response presented during the development of this methodology, identified currently as adherent to the cartography in health care. Previously, the methodology was called only of FAO and was used centesimal, millesimal, and 50th millesimal scales throughout the history of its development [38–41].

To reach to its innovative state of the art, we note and compared practical concepts of assessment of the biofield described traditionally by Traditional Chinese Medicine, Vitalism, and others; by consensual centers of awareness, traditionally known as chakras by Ayurveda, from India, which inspired to define the homeopathic medicines related to each part of the human body represented in seven endocrine centers; by sequential order of use of each one inspired in the traditional knowledge of the alchemy; by an amount of repetitions inspired by the traditional knowledge of the sacred geometry; and by concepts of the quantum physics [39, 40].

3. Material and methods

The BioFAO methodology was developed by observation of effects of clinical treatment adherents to the cartography in health care methodology. The medicines that are part of BioFAO are *Antimonium crudum, Kali carbonicum, Mercurius solubilis, sulfur, Natrum muriaticum, Aurum metallicum,* and *Ammonium muriaticum*. Currently, the matrix used has the aliases Archeus and uses repetitions of these medicines according to described in the publication of the methodology [39, 40]. Each one of these medicines has inspired connection from consensual centers of awareness that related with the endocrine centers, described in **Table 1**.

Each cycle of administration of the therapeutic methodology there are used four (4) bottles with interval of one (1) hour between them in the same day. Each bottle has a complex with 25 dosages of the seven mineral medicines in two different dynamizations. Therefore, after the administration of the 4 bottles, 100 dosages are used in a total of one (1) cycle of protocol. This information only should be used for a description of a methodology, and should be prescribed and monitored by a qualified and updated professional in specific training course of the BioFAO Institute. The medicine described to be applied for use, either for clinical treatment or for research, should be prepared by accredited facility of reference Pharma Lotus and Alquiotupã Pharmacy. The self-medication is inadvisable in any case, even as administer the medicine to a person or animal with no indication or authorization of the doctor responsible, updated, and qualified to assess the monitoring of the effects [38, 42, 43].

Homeopathic medicine	Endocrine centers
Antimonium crudum	Pituitary gland
Kali carbonicum	Pineal gland
Mercurius solubilis	Thyroid and parathyroid glands
Sulfur	Pancreas
Natrum muriaticum	Thymus gland
Aurum metallicum	Gonads
Ammonium muriaticum	Adrenal

Table 1.
List of the medicines used and its correspondences to the endocrine centers of the region of thematic influence.

4. Key results

4.1 Homeopathic aspects of Self-Organizing Factors of Biofield

The discovery of the BioFAO medicines was different from proving, how it is described in the classic Homeopathy with the use in repeated and systematized doses. The clinical treatment of an essentially rural population considered healthy in global way supported the development and adjustments of the BioFAO methodology whenever necessary. The trajectory of patients of the office of a doctor's group working professionally in Rio de Janeiro's city was used as comparative elements, and, yet, a restrict group spontaneously intoxicated by pesticides with health status considered incurable and chronic. The pictures that appear in these distinct situations were monitored according to the symptomatology, the clinical resolution, and the potencies used, with discussion mainly in an oral form of the results and adjustments in the treatments. Usually, potency bands are assigned according to the organic injury and the case individually, but the set of some cases specifically could be presented by pathology. These specific cases eventually share a minimum or maximum potency of general suggestion, but the individual response continues to predominate in the assessment. La Gioconda effect kept together patients, doctors, and pharmacist team around the desire to be connected to the treatment with the BioFAO methodology during almost 40 years of its development.

After the beginning of a treatment with BioFAO, it is reported that the patient triggers a healing process because they present phenomena compatible with the Hering's Healing Laws. Moreover, well-described stages are identified in the book in Portuguese, known traditionally by alchemy as albedo, nigredo, and rubedo. During the healing process are described unique events, incidents, and situations related to prior decisions, behavior, and choices, but that reoccur to grant the possibility to be reviewed, reconsidered, and redirected. The full experience is identified by author herself as synchronicity, which refers subjectively to a metaphor of spiritual connection of the life phenomenon represented in the individual at issue with his connection to the concept of force of the universe. It is also noted phenomena are compatible with the subjective feeling of general well-being, commonly described when a person uses the medicine and the potency that would be his *Simillimum* by the traditional knowledge of the Homeopathy. The biological aspect equivalent to this phenomenon in the living being is compatible with self-regulation in the immune system and happens a balance, discovery, or indication of the healing path of the physical symptoms [39, 40, 44].

The potencies used in 2018 are between 9 and 15d and their use varies according to the history obtained in the anamnesis of the patient. The medicines are complex according to the Brazilian Homeopathic Pharmacopeia and the administration happens in 4 (four) dosages with an interval of 1 h between them. Generically, the chronic and injury cases use more low potencies, acute and functional cases use more high potencies. This protocol of global antipsoric action is repeated, in average, at each 3 months, but varied according to the case and data obtained in the anamnesis of the observations reported by patient or guardian.

The potency indication is much more related to the particularity of the individual than exclusively to the health-disease process, as it is common to believe. Some notes can indicate that situations of illness can be necessary specific potencies, but it is not the general rule to use of the BioFAO methodology. This dynamic is very well explained and taught in the training course and with the clinical practice to use this complex of dynamized medicines. Contrary to the criteria defined for humans, the plants use the centesimal scale and fixed potencies, with results presented in a material with reports of several experiments performed in Brazil [39, 45].

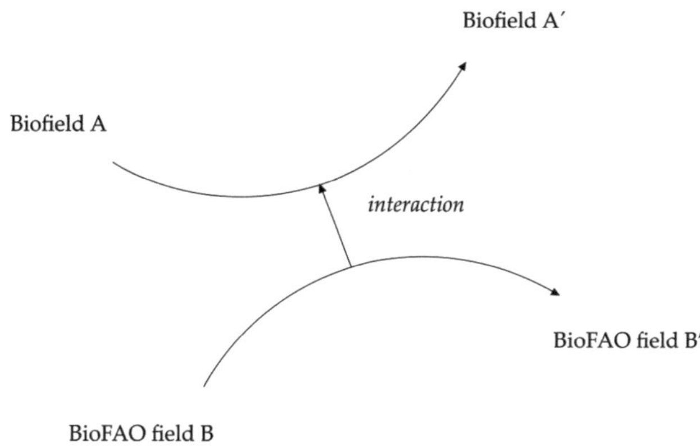

Figure 2.
Proposal of representation of the interaction of the BioFAO field with the biofield through the mediator virtual photon described by Penrose [49]. The biofield of the living organism in the status A change in the status A' after the administration of the BioFAO.

The use of the BioFAO methodology with animals came from the development with humans. The translation of the evolution languages and the severity of the pathologies should be considered according to each animal, which also can vary between the species. Therefore, the veterinarian should do the indication considering the reality in the moment of the patient and the impact and depth that the health-disease processes are in course. However, the use of dynamized medicines with animals that were initially developed for humans happened historically with some other styles *lato sensu* in the Homeopathy. There is a simplified proposal that uses the centesimal scale for animals, but with specific restrictions to some species and age group [46–48].

The interaction of the medicine field with the biofield is a subject of high curiosity because, qualitatively, there is a possible change of state of the biofield after the exposure to medicine (**Figure 2**). It is known that there is the transfer of photons in the interaction between fields, according to the virtual photon as a mediator of the interaction described by Penrose. This phenomenon is well described in the quantum physics; it is expected to understand the complexity of interaction of the biofield with field composed by 100 dosages of 7 homeopathic medicines from dynamized minerals [39, 49].

4.2 Obstacles to the healing

There are observed interference elements by physical and chemical agents in the treatment with BioFAO. The most common physical agents that change the effect of the medicine, before to be used in the treatment, are: heat sources, X-rays, and exposure to electromagnetic waves. The most common chemical agents in the interference in the living organisms, after receiving the treatment, are mainly when exposed to: camphor, menthol, and eucalyptol.

About the physical agents, their interaction motivates to believe in the possibility that information of the medicine is coded in a field due to the heat sensitivity or the strong electromagnetic field agents sensitivity. Particularly, they are noted when a concern to the BioFAO complex that has only dynamized active principles. The examples of remarkable situations in which the BioFAO seeming to have received destructive interference in their effects, happened by demonstrations of controlled researches with DL50 in poisoning of animal models, and by empirical observations of the cartography in health care of the clinical treatments with humans. Authors estimate that the homeopathic medicines present weak fields, so, they would be sensitive to heat and to some radiation sources. These findings present interference agents in the homeopathic medicines from well-known elements in the Physics area [50, 51]. Let us see practical examples:

The typical example of exposure to heat is to keep the medicine in the car parked in the sun. The internal temperature of the vehicle easily can double regarding to ambient temperature, external to the vehicle, which works like a greenhouse. In a city of tropical weather, like Rio de Janeiro, the ambient temperature easily reaches 30°C (86°F) during the summer. Therefore, the internal temperature of a vehicle parked in the sun easily reaches 50°C (122°F). The exposure of dynamized medicines to temperature equal or higher to 50°C (122°F) has capacity to inactivate and, therefore, avoid the correct functioning of the homeopathic medicine. This type of interference can be simulated by a laboratory heat incubator. The inactivation of dynamized medicines by heat is described in the literature, used as a way of disinformation of impregnated containers and noted in experimental protocols. The solution to preserve the medicine in risk of heat inactivation in a vehicle is to remove from car while is parked, so, a simple option is carry it with you [52, 53].

Expose to electromagnetic fields of some electronic equipment, like cell phone during the battery charge cycle or personal computers for 15 min, at a distance of up to 1 m (3.3 feet) inactivates the correct functioning of a BioFAO medicine. The safety distance noted as minimum recommended for the main usual household equipment is 2 m (6.6 ft). Other electronic equipment with potentially harmful electromagnetic field can be present in a situation of temporary storage during the service of home delivery. The entrance of a building with security technology usually has computers and electronics of access monitoring by remote surveillance cameras, and personal cell phones of the staff in battery charge during the working hours. The place of temporary storage of orders and mails usually is less than 2 m (6.6 ft) of this equipment and interfere deleteriously in the effects of the dynamized medicines BioFAO. However, if the medicine is in a fully sealed aluminum packaging, the protection prevents that exposure causes effects that harm the functioning of the BioFAO medicine [50].

Another example would be exposing of the medicine to X-rays, like what happens in airports and in the postal service, in the equipment used in the inspection of baggage or cargo. There is controversial literature about the exposure influence to the X-rays as an interference source in ultradiluted medicines. To the BioFAO complex, there were harmful interferences in the expected effects of the treatment protocols. The findings of these deleterious effects justified special care for transport in travels with BioFAO medicine. One way to avoid this deleterious exposure is to carry the medicine in the hand luggage with a document translated containing a statement that the passenger is in medical treatment and carries a homeopathic medicine that should be inspected without the use of X-rays in the usual hand luggage check. Another option, with a functional effect that avoids such interference, is to transport the medicine inside a lead bag (made with a malleable material with lead in its structure), the same used for protection of photo negative films [51].

The living organism, under influence of the medicine, also is susceptible to suffer interference, but, in this case, the agents are a little different of those that affect the medicine. Observing the personal reports repeated throughout the years can occur unbalance of the organism with reappearance of symptoms, as it is seemingly before to start the treatment, which are different of reappearance of old symptoms, as described by Healing Laws. Besides, the change in the experiences of synchronicity reported is remarkable. The exposure of a person treated with BioFAO to computed tomography interrupts the BioFAO effect, but, seemingly, the exposure of a person to magnetic resonance, ultrasonography, cell phone charging, or use of electronic equipment does not have this interference capacity. What happens in the person is the mechanism of action of the tomography while the physical agent of interference in the organism on the effect of BioFAO methodology remains unknown and opens an innovative research field. The solution would be to repeat the protocol after and exam of computed tomography. The physical agents have minor importance in the interference of the treatment, but, it is not possible to say it of the chemical factors, as we can note.

Specific substances have to be monitored by own person to avoid exposure during his treatment or of an animal, while it is desired to maintain the effects of the BioFAO methodology. These data were obtained from objective reports of patients and guardians that noted the effects interrupted suddenly. During the anamnesis of the appointment, after a "break" of the effect, appears in the history of exposure nasal path, possibly olfactory, or oral, to the chemical substances camphor, eucalyptol, and/or menthol. Besides these substances identified, the exposure to the mothball scent, some routine vaccines, surgical procedures, and the oral treatment with corticosteroids in the dosage recommended of the hormone, like anti-inflammatories should be monitored with attention [54].

These data open a perspective that a change in the hormonal axis of the stress can be strong enough to interfere in the biofield of the organism with the BioFAO. It is known that this hormonal chain is interconnected to several processes of physiological self-regulation and immunity, but it is not precisely known how these interferences change the effect of the BioFAO methodology in the biofield.

The traditional Hahnemannian literature corroborates some of these facts identified, such as the effect that camphor is able to neutralize the functioning of homeopathic medicines. In parallel, a stressful event for a person can trigger the identification by patient and report spontaneously to the doctor the need of new medicine. So, when exposed to one of these agents, the experienced patient learns to note and identify when is necessary new administration, and reports it to the doctor. Currently, these substances are found in many industrialized products, especially foods with flavor, personal hygiene, and cleaning products. Many times they are found in the labels inaccurately or generically as flavor, which conceals the real concentration of the taste and the substance present and the careful patient only discovery when exposed (interrupting his treatment). This situation can be bypassed requesting to a person that does not use BioFAO to identify the product characteristics, like, for example, a person or a seller. However, not all people know to identify correctly the smell and the taste of camphor, eucalyptus, and mint [10, 39, 40, 54].

The most known reference products that have these components in high concentration are bruise sprays usually used after sport injuries without a prescription, plasters, unguent, a traditional bronchodilator in ointment, mint candies, candies that chill the mouth, cleaning products with odor of eucalyptus, essence of eucalyptus in sauna, solutions for mouthwash, and in variable concentrations of toothpaste, dental floss. It is apparently challenging to request to avoid the exposure to these agents, but it is perfectly possible to live without them. At the same time, it is curious as arise the reports of peculiar and unexpected exposures, usually in the first uses of the treatment, like, for example, exposure to camphor in the bathroom trash can or use of eucalyptus essence in the toilet to conceal the bad fecal odor. Most of the products that provide exposures to these substances have substitutive or different solutions from original proposal, without the use of none of the agents mentioned. Eventually, urinals in male bathrooms use naphthalene (bookworm repellent in clothes and book cabinets) to conceal bad smell of urine. Researchers suggest in several papers to abolish the use of these substances, although without knowing its context in the interaction with BioFAO [10, 39].

5. General discussion

Several elements of discussion were presented throughout the results. The trajectory of Mona Lisa and the Homeopathy presents remarkable historical and common aspects, among them. The description of the *La Gioconda* effect is peculiar regarding to the way with which the facts were presented in the Homeopathy. Both are example of a very strong interest over people due to be very attractive in the contemporaneity. Eventually, there are different opinions about them, which is healthy to the development of any area. Movements and statements are noted in favor and contrary Homeopathy, sometimes by fragments of the general society, others specifically by fragments of academic groups of other areas, but that not necessarily covers the total opinion of society, population, or specialized researchers [1, 2, 4–9, 11–15].

The evolution of the scientific knowledge happens step by step, in sometimes the observation of the natural phenomenon is more advanced, in other times there is the emergence of coherent theories to explain the phenomenology, so, theory and practice coevolve. In the case of Homeopathy, the set of natural phenomenon observable by use of dynamized medicines is more advanced that the full theoretical consolidation. The explanation by theories that serve for parts of what happens in the Homeopathy, although is known how to use the therapeutic tool in clinical way, eventually arouses in the people hearty movements by small groups that eventually are very announced by local media. By the other side, a part of the population of rural producers in Brazil identified advantages, defend, and use homeopathic medicines to the farm animals, for all benefits that consider be provided in the management of findings [7–9, 11–14, 32–41, 43, 44, 46, 47].

Notoriously, as a general discussion, the interested people to keep under effect of the BioFAO treatment contact the manufacturers of products to confirm the presence of interference agents through the consumer service. Therefore, they track their products more used in the day by day and discover compatible substitutes. From this point of view, we are faced with seemingly unnecessary substances that people and animals are exposed, sometimes more than three times a day; however, without knowing exactly what is your real need and the adverse effects in the organisms, in the short and/or long term. These substances should be mandatorily informed when present in all labels of the industrialized products, for the user can choose and decide to which exposes. Once again, the *La Gioconda* effect currently moves a legacy around 15,000 patients and guardians imbued in the desire to maintain actively with the functional BioFAO treatment and their self-regulation, being 10,000 patients in Rio de Janeiro and 5,000 outside Rio de janeiro [43, 54].

We can identify the personal bound that was put in the several subjects described throughout the chapter, such as the Renaissance for Leonardo da Vinci, the classic Homeopathy for Samuel Hahnemann, the anthroposophy for Rudolf Steiner, and BioFAO for Miria de Amorim. It is expected that the legacy of the Homeopathy offered to the plants, animals, and people can maintain in continuous update to serve to its interested public, coherent with the claims, and challenges of the society throughout the time [1, 9–12, 31, 34, 35, 39, 40, 45, 46].

6. Conclusions

People show personal and professional attraction by Homeopathy, even as by effects of dynamized or ultradiluted medicines.

There are qualitative similarities between the description of the history of Mona Lisa and the history of the Homeopathy.

The attraction of the people by Homeopathy is described as *La Gioconda* effect that maintains alive the development of the Healing Art.

The use of the word Homeopathy requests flexibility due to do the continuous growth of this area. The contemporaneity brought the demand of *lato sensu* sense of the word Homeopathy to aggregate the derived knowledge that uses dynamized medicines from Homeopathy in the *stricto sensu*, in other words, the Hahnemannian Classical Homeopathy.

The methodology self-organizing factors of the biofield (BioFAO) uses dynamized medicines, inspires in traditional knowledge, and provides compatible phenomena with the purposes of the Homeopathy.

BioFAO with animals evolves from learning with the development of the methodology from experiences with humans.

Acknowledgements

Natasha Zadorosny Lopes Bastos who translated from Portuguese to English. Instituto BioFAO for general support.

Conflict of interest

The author has no conflict of interest to declare.

Notes

The images used are digitalization of works of art with more than 100 years; the Adobe Photoshop free trial (c) software used for the edition of images was licensed during the period of its free validity.

Acronyms and Abbreviations

BioFAO	Self-Organizing Factors of Biofield
CT ou CAT scans	Computed tomography scan
SWL	shock wave lithotripsy

Appendices and Nomenclature

c	centesimal hahnemannian dynamization (potentisation) scale
d	decimal hahnemannian dynamization (potentisation) scale
Wikipedia	free online encyclopedia

Author details

Silvio Leite Monteiro da Silva
Universidade Federal do Estado do Rio de Janeiro-UNIRIO (Federal University of State of Rio de Janeiro), Rio de Janeiro, Brazil

*Address all correspondence to: silvioleite@gmail.com

References

[1] Scailliérez C, La Gioconda. France: Louvre Museum; 2018. Available from: https://www.louvre.fr/en/mediaimages/portrait-de-lisa-gherardini-epouse-de-francesco-del-giocondo-dite-monna-lisa-la-gioconda [Accessed: September 9, 2018]

[2] Zelazko A. Why is the Mona Lisa so famous? In: Britannica. 2018. Available from: https://www.britannica.com/story/why-is-the-mona-lisa-so-famous [Accessed: September 9, 2018]

[3] The French Ministry of Foreign Affairs. The Louvre, The Most Visited Museum in the World. France Diplomatie: Ministry of Foreign Affairs and International Development; 2015. Available from: https://www.diplomatie.gouv.fr/en/french-foreign-policy/tourism/events/article/the-louvre-the-most-visited-museum [Accessed: October 9, 2018]

[4] Moraes ED. Mona Lisa: Multiple meanings of an enigmatic smile. DELTA. 2013;**29**(Especial):443-465. Available from: http://www.scielo.br/pdf/delta/v29nspe/v29nspea05.pdf ISSN 0102-4450

[5] Guinness World Records. Highest insurance valuation for a painting. [Accessed: October 9, 2018]. Available from: http://www.guinnessworldrecords.com/world-records/highest-insurance-valuation-for-a-painting

[6] Asmus JF, Parfenov VA, Elford JP. Seeing double: Leonardo's Mona Lisa twin. Optical and Quantum Electronics, Springer US. 2016;**48**(12):1-11. DOI: 10.1007/s11082-016-0799-0

[7] Relton C, Cooper K, Viksveen P, Fibert P, Thomas K. Prevalence of homeopathy use by the general population worldwide: A systematic review. Homeopathy, Elsevier Ltd. 2017;**106**(2):69-78. DOI: 10.1016/j.homp.2017.03.002

[8] The Lancet. The end of homoeopathy. Lancet [Editorial]. 2005 Aug;**366**(9487):690. Available from: http://linkinghub.elsevier.com/retrieve/pii/S0140673605671498

[9] Dessaix JM. L'Omepatie e le Sue Prime Bataglie. Introduzione, Traduzione e Note di Prof. Paolo Bellavite. Salus Infirmorum: Padova; 2013. 104 p. Available from: http://paolobellavite.it/files/OmeoBattaglie_f4klk3ce.pdf

[10] Hahnemann S. Organon of the Rational Art of Healing. London: J. M. Dent & Sons, Ltd. New York: E. P. Button & Co.; 1913. 244 p. Available from: https://archive.org/details/organonofrationa00hahn

[11] Waisse S. Private and institutionalised patients' use of homeopathy in the early nineteenth century. Homeopathy. 2017;**106**(4):250-259. Available from: http://www.sciencedirect.com/science/article/pii/S1475491617300486. DOI: 10.1016/j.homp.2017.07.001

[12] Loudon I. A brief history of homeopathy. The Royal Society of Medicine. 2006;**99**(12):607-610. Available from: http://www.ncbi.nlm.nih.gov/pmc/articles/PMC1676328/

[13] Ullman D. Dysfunction at Wikipedia on Homeopathic Medicine. 2014; 2014-10-10. Available from: https://www.huffingtonpost.com/dana-ullman/dysfunction-at-wikipedia-_b_5924226.html [Accessed: September 11, 2018]

[14] Ehrenfreund M. The Science of Wikipedia Flamewars. 2013; (July 23, 2013). Available from: https://www.washingtonpost.

com/news/wonk/wp/2013/07/23/the-science-of-wikipedia-flamewars/?noredirect=on&utm_term=.89f3799f6b99 [Accessed: September 11, 2018]

[15] Ehrenfreund M. The Moving of the Mona Lisa. The Independent. 2005. Saturday 2 April. Available from: https://www.independent.co.uk/news/world/europe/the-moving-of-the-mona-lisa-530771.html [Accessed: September 14, 2018]

[16] Soroko M, Howell K. Infrared thermography: Current applications in equine medicine. Journal of Equine Veterinary Science. [Internet]. 2018;**60**:90-96.e2. Available from: https://doi.org/10.1016/j.jevs.2016.11.002

[17] Swartz BE. The advantages of digital over analog recording techniques. Electroencephalogr Clin Neurophysiol. 1998;**106**(2):113-117

[18] The Nobel Prize in Physics 1901. NobelPrize.org. Nobel Media AB 2018. Available from: https://www.nobelprize.org/prizes/physics/1901/summary/ [Accessed: September 10, 2018]

[19] The Nobel Prize in Physiology or Medicine 1924. NobelPrize.org. Nobel Media AB 2018. Available from: https://www.nobelprize.org/prizes/medicine/1924/summary/ [Accessed: September 10, 2018]

[20] The Nobel Prize in Physiology or Medicine 1979. NobelPrize.org. Nobel Media AB 2018. Available from: https://www.nobelprize.org/prizes/medicine/1979/summary/ [Accessed: September 10, 2018]

[21] The Nobel Prize in Physiology or Medicine 2003. NobelPrize.org. Nobel Media AB 2018. Available from: https://www.nobelprize.org/prizes/medicine/2003/summary/ [Accessed: September 10, 2018]

[22] Pinna M, Manchia M, Oppo R, Scano F, Pillai G, Loche AP, et al. Clinical and biological predictors of response to electroconvulsive therapy (ECT): A review. Neuroscience Letters. 2018;**669**:32-42. DOI: 10.1016/j.neulet.2016.10.047

[23] Leary S, Underwood W, Anthony R, Cartner S. AVMA guidelines for the euthanasia of animals: 2013 edition [Internet]. 2013th ed. American Veterinary Medical Association. 2013. 98 p. Available from: https://www.avma.org/kb/policies/documents/euthanasia.pdf

[24] Jones GRN. Judicial electrocution and the prison doctor. Lancet. 1990;**335**(8691):713-714. Available from: http://www.sciencedirect.com/science/article/pii/014067369090816N

[25] Rault E, Lacornerie T, Dang H-P, Crop F, Lartigau E, Reynaert N, et al. Accelerated partial breast irradiation using robotic radiotherapy: A dosimetric comparison with tomotherapy and three-dimensional conformal radiotherapy. Radiation Oncology. 2016;**11**(1):29. Available from: http://www.ro-journal.com/content/11/1/29. DOI: 10.1186/s13014-016-0607-9

[26] Cantillon DJ, Dukkipati SR, Ip JH, Exner DV, Niazi IK, Banker RS, et al. Comparative study of acute and mid-term complications with leadless and transvenous cardiac pacemakers. Heart Rhythm. 2018;**15**(7):1023-1030. DOI: 10.1016/j.hrthm.2018.04.022

[27] Pourmand A, Galvis J, Yamane D. The controversial role of dual sequential defibrillation in shockable cardiac arrest. The American Journal of Emergency Medicine. 2018;**36**(9):1674-1679. DOI: 10.1016/j.ajem.2018.05.078

[28] Sorensen MD, Bailey MR, Hsi RS, Cunitz BW, Simon JC, Wang YN, et al. Focused ultrasonic propulsion of kidney stones: Review and update of preclinical

technology. Journal of Endourology. 2013;**27**(10):1183-1186. DOI: 10.1089/end.2013.0315

[29] Dutra-Correa M, Nicolodelli G, Rodrigues JR, Kurachi C, Bagnato VS. Femtosecond laser ablation on dental hard tissues—Analysis of ablated profile near an interface using local effective intensity. Laser Physics. 2011;**21**(5):965-971. DOI: 10.1134/S1054660X11090064

[30] Bingen BO, Engels MC, Schalij MJ, Jangsangthong W, Neshati Z, Feola I, et al. Light-induced termination of spiral wave arrhythmias by optogenetic engineering of atrial cardiomyocytes. Cardiovascular Research. 2014;**104**(1):194-205. DOI: 10.1093/cvr/cvu179

[31] Caulfield KA, George MS. The future of brain stimulation treatments. The Psychiatric Clinics of North America. 2018;**41**(3):515-533. DOI: 10.1016/j.psc.2018.05.004

[32] Holandino C, Oliveira AP, Homsani F, de Paiva JP, Barbosa GM, de Lima Zanetti MR, et al. Structural and thermal analyses of zinc and lactose in homeopathic triturated systems. Homeopathy. 2017;**106**(3):160-170. DOI: 10.1016/j.homp.2017.06.003

[33] Bellavite P, Marzotto M, Olioso D, Moratti E, Conforti A. High-dilution effects revisited. 1. Physicochemical aspects. Homeopathy. 2014;**103**(1):4-21. DOI: 10.1016/j.homp.2013.08.003

[34] Barthel P. Hahnemann's legacy-the Q (LM) potencies. The British Homoeopathic Journal. 1991;**80**(2):112-121. DOI: 10.1016/S0007-0785(05)80395-3

[35] Bonamin LV, De Carvalho AC, Waisse S. *Viscum album* (L.) in experimental animal tumors: A meta-analysis. Experimental and Therapeutic Medicine. 2017;**13**(6):2723-2740. DOI: 10.3892/etm.2017.4372

[36] Teixeira MZ. New homeopathic medicines' database: A project to employ conventional drugs according to the homeopathic method of treatment. The European Journal of Integrative Medicine. 2013;**5**(3):270-278. DOI: 10.1016/j.eujim.2013.01.001

[37] Bonamin LV, Endler PC. Animal models for studying homeopathy and high dilutions: Conceptual critical review. Homeopathy. 2010;**99**(1):37-50. DOI: 10.1016/j.homp.2009.11.003

[38] Kharal A. Homeopathic drug selection using intuitionistic fuzzy sets. Homeopathy. 2009;**98**(1):35-39. DOI: 10.1016/j.homp.2008.10.003

[39] Amorim M. BioFAO: Uma Medicina Para Um Novo Tempo. 1st ed. Rio de Janeiro: Editora Boneker; 2018. 364 p. Available from: https://www.lojabonecker.com.br/biofao-uma-medicina-para-um-novo-tempo

[40] Amorim M. Holismo, Homeopatia, Alquimia Uma Sincroniciade Para a Cura. 1st ed. Rio de Janeiro: Editora Mauad; 2000. 240 p. Available from: http://mauad.com.br/index.php?route=product/product&filter_name=sincronicidade&filter_description=true&filter_sub_category=true&product_id=13488

[41] Ferigato SH, Carvalho SR. Pesquisa qualitativa, cartografia e saúde: Conexões . Vol. 15, Interface - Comunicação, Saúde, Educação, Scielo; 2011. p. 663-676

[42] Instituto BioFAO. Available from: http://www.institutobiofao.org.br [Accessed: October 15, 2017]

[43] Farmácia Alquiotupã. Available from: http://www.alquiotupa.com.br [Accessed: October 15, 2017]

[44] Vaz NM, Carvalho CR. On the origin of immunopathology. Journal of Theoretical Biology. 2015;**375**:61-70. DOI: 10.1016/j.jtbi.2014.06.006

[45] Ranal MA, Amorim M. BioFAO na Agricultura: Recuperação da Defesa Natural das Plantas. 1st ed. Rio de Janeiro: Editora Bonecker; 2017. 227 p. Available from: https://www.lojabonecker.com.br/biofao-na-agricultura-recuperacao-da-defesa-natural-das-plantas-

[46] Monteiro da Silva SL. Complementary medicine with high dilutions strengthen conventional therapies and health. In: Kukovics S, editor. Goat Science. Rijeka: IntechOpen; 2018. DOI: 10.5772/intechopen.72415

[47] Moreira HM, Amorim M, Maruyama C, Trinca R, Torres C, Ornellas R, et al. Survival of mice with Erhlich ascitic tumour treated with ultra-dilutions. EJC. 2011;**47**(1):S100

[48] Lacerda EB, de Almeida Rezende Machado NV, de Souza GH, Bonamin L, Bernardi MM, Monteiro da Silva SL. Dairy goats endoparasites infection and effects of commercial homeopathy medicine. International Journal of High Dilution Research. 2013;**12**(44):156-157

[49] Penrose R. A Mente Nova Do Rei - Computadores, Mentes, e as Leis da física. [The Emperos's New Mind]. 2nd ed. Rio de Janeiro: Editora Campus; 1993. 466 p

[50] Weber S, Endler PC, Welles SU, Suanjak-Traidl E, Scherer-Pongratz W, Frass M, et al. The effect of homeopathically prepared thyroxine on highland frogs: Influence of electromagnetic fields. Homeopathy. 2008;**97**(1):3-9. DOI: 10.1016/j.homp.2007.11.002

[51] Moreira HM, Amorim M, Ornelas RH, Bignardi F, Moreira VO, Guimarães APC, Yasuda TA, Pacheco H. Um estudo demonstrando que a passagem do medicamento homeopático pelo aparelho de raio x modifica sua ação produzindo efeitos deletérios nos mesmos. Rio de Janeiro: Instituto BioFAO;2017. Available from: http://institutobiofao.org.br/wp-content/uploads/2017/09/UM-ESTUDO-DEMONSTRANDO-QUE-A-PASSAGEM-DO-MEDICAMENTO-HOMEOPÁTICO-PELO-APARELHO-DE-RAIO-X-MODIFICA-SUA-AÇÃO-PRODUZINDO-EFEITOS-DELETÉRIOS-NOS-MESMOS.pdf [Accessed: September 11, 2018]

[52] Gibbs LI, Lawrence DW, Kohn MA. Heat exposure in an enclosed automobile. Journal of the Louisiana State Medical Society. Dec 1995;**147**(12):545-546

[53] McLaren C. Heat stress from enclosed vehicles: Moderate ambient temperatures cause significant temperature rise in enclosed vehicles. Pediatrics [Internet]. 2005;**116**(1):e109-e112. Available from: http://pediatrics.aappublications.org/cgi/doi/10.1542/peds.2004-2368

[54] Manoguerra AS, Erdman AR, Wax PM, Nelson LS, Martin Caravati E, Cobaugh DJ, et al. Camphor poisoning: An evidence-based practice guideline for out-of-hospital management. Clinical Toxicology, Taylor & Francis. 2006;**44**(4):357-370. Available from: https://doi.org/10.1080/15563650600671696

Chapter 5

A Review on Natural Antioxidants

Arun Rasheed and Rinshana Fathima Abdul Azeez

Abstract

Free radicals and related species have attracted a great deal of attention in recent years. Oxidative stress has been considered a major contributory factor to the diseases. They are mainly derived from oxygen (reactive oxygen species (ROS)) and nitrogen (reactive nitrogen species (RNS)) and are generated in our body by various endogenous systems and exposure to different physicochemical conditions or pathophysiological states. Free radical damage to protein can result in loss of enzyme activity. There are epidemiological evidences correlating higher intake of components/foods with antioxidant abilities to lower incidence of various human morbidities or mortalities. The sources and origin of antioxidants which include fruits and vegetables, meats, poultry, and fish were treated in this study. The classification and characteristics of antioxidant, its measurements and level in food and free radicals, were also documented. The chemistry of antioxidants which includes chain reactions, molecular structures, food antioxidants and reaction mechanisms, biochemical activity, therapeutic properties, and future choice of antioxidants was reported in this review.

Keywords: antioxidants, free radicals, oxidative stress

1. Introduction

Plants such as shrubs, herbs, or trees in parts or in whole were used in the treatment and management of various diseases, and disorders can be dated long back. Natural phytochemicals present at low levels in fruits, vegetables, herbs, and spices offer many health benefits, but these compounds may not be effective or safe when consumed at higher dose [1]. The presence of free radicals in biological materials was discovered less than 50 years ago [2].

Pollutants, ionizing radiation or UV light, smoking, exposure of biological systems to xenobiotics, and development of certain pathological conditions lead to oxidative stress, thereby increases production of oxy radicals [3]. Cell damage caused by free radicals appears to be a major contributor in aging and degenerative diseases such as cancer, cardiovascular disease, cataracts, rheumatoid arthritis, and brain dysfunction. Free radicals have been implicated in the pathogenesis of at least 50 diseases. Fortunately, free radical formation is controlled naturally by various beneficial compounds and antioxidants, and its availability is limited that this damage can become cumulative and debilitating. Antioxidants are capable of stabilizing, deactivating, or scavenging free radicals before they attack cells.

Reactive species	Symbol	Half-life (in seconds)	Reactivity/remarks
Reactive oxygen species			
Superoxide	$O_2^{\cdot}$	10^{-6} s	Generated in mitochondria, in cardiovascular system, and others

Hydroxyl radicle	$^{\cdot}OH$	10^{-9} s	Very highly reactive, generated during iron overload and such conditions in our body
Hydrogen peroxide	H_2O_2	Stable	Formed in our body by a large number of reactions and yields potent species like. OH
Peroxyl radical	ROO*	S	Reactive and formed from lipids, proteins, DNA, sugars, etc. during oxidative damage
Organic hydroxide	ROOH	Stable	Reactive with transient metal ions to yield reactive species
Singlet oxygen	$^1O_2$	10^{-6} s	Highly reactive, formed during photosensitization and chemical reactions
Ozone	O_3	S	Present as an atmospheric pollutant can react with various molecules
Reactive nitrogen species			
Nitric oxide	NO*	S	Neurotransmitter and blood pressure regulator can yield potent oxidants during pathological status
Peroxy nitrile	$ONOO^-$	10^{-3} s	Formed from nitric oxide and superoxide and highly reactive
Peroxynitrous acid	ONOOH	Fairly stable	Protonated from of $ONOO^-$
Nitrogen dioxide	NO_2	S	Formed during atmospheric pollution

Antioxidants can be defined as substances whose presence in relatively low concentrations significantly inhibits the role of oxidation of the targets. Due to continuous generation of partially reduced forms of oxygen by constitutive metabolic pathways, a number of protective antioxidant enzymes, such as superoxide dismutase (SOD), catalase (CAT), glutathione peroxidase (GSHPx), glutathione reductase (GSHRx), glutathione-S-transferase (GST), and nonenzymatic antioxidants, have involved to deal with toxic species. Oxidation is a chemical reaction that transfers electrons from a substance to an oxidizing agent. Oxidation reactions can produce free radicals, which start chain reactions that damage cells. Antioxidants terminate these chain reactions by removing free radical intermediates and inhibit other oxidation reactions by being oxidized themselves. Antioxidants are often reducing agents such as thiols, ascorbic acid, or polyphenols.

1.1 Sources and origin of antioxidants

Antioxidants are abundant in fruits and vegetables, as well as in other foods including nuts, grains, and some meats, poultry, and fish. β-Carotene is found in many foods, including sweet potatoes, carrots, cantaloupe, squash, apricots, pumpkin, and mangoes. Lutein, best known for its association with healthy eyes, is abundant in green, leafy vegetables such as collard greens, spinach, and kale. Lycopene is a potent antioxidant found in tomatoes, watermelon, guava, papaya, apricots, pink grapefruit, blood oranges, and other foods. Estimates suggest 85% of American dietary intake of lycopene comes from tomatoes and tomato products [4].

1.1.1 Types of antioxidants

Antioxidants are grouped into two:

1. Primary or natural antioxidants

2. Secondary or synthetic antioxidants

1.1.1.1 Primary or natural antioxidants

They are the chain breaking antioxidants which react with lipid radicals and convert them into more stable products. They are mainly phenolic in structures and include the following [5]:

1. Antioxidant minerals: These are cofactor of antioxidants enzymes. Their absence will definitely affect metabolism of many macromolecules such as carbohydrates. Examples include selenium, copper, iron, etc.

2. Antioxidant vitamins: They are needed for most body metabolic functions. They include vitamin C, vitamin E, and vitamin B.

3. Phytochemicals: These are phenolic compounds that are neither vitamins nor minerals. These include:

Flavonoids: These are phenolic compounds that give vegetables fruits, grains, seeds leaves, flowers, and bark their colors. Catechins are the most active antioxidants in green and black tea and sesamol. Carotenoids are fat soluble color in fruits and vegetables. Zeaxanthin is high in spinach and other dark greens.

1.1.1.2 Secondary or synthetic antioxidants

These are phenolic compounds that perform the function of capturing free radicals and stopping the chain reactions; the compound includes [5]:

1. Butylated hydroxyanisole (BHA)

2. Butylated hydroxytoluene (BHT)

3. Propyl gallate (PG) and metal chelating agent (EDTA)

4. Tertiary butylhydroquinone (TBHQ)

5. Nordihydroguaiaretic acid (NDGA).

2. Classification

- **Enzymatic antioxidants:**

  1. Primary antioxidants, for example, SOD, catalase, glutathione peroxidase

  2. Secondary enzymes, for example, glutathione reductase, glucose-6-phosphate dehydrogenase

- **Nonenzymatic antioxidants:**

  1. Minerals, for example, zinc, selenium

  2. Vitamins, for example, vitamin A, vitamin C, vitamin E

  3. Carotenoids, for example, β-carotene, lycopene, lutein, zeaxanthin

4. Low-molecular weight antioxidants, for example, glutathione, uric acid

5. Organosulfur compounds, for example, allium, allyl sulfide, indoles

6. Antioxidant cofactors

7. Polyphenols

2.1 Enzymatic antioxidants

2.1.1 Copper/zinc and manganese dependent

Superoxide dismutase (SOD): SOD is a group of endogenously produced metalloenzymes with various prosthetic groups present both in prokaryotes and eukaryotes [6]. Three main classes of them differ in their amino acid sequence structure and metallic factors as follows:

1. Cu-Zinc SOD in the cytoplasm with two sub-units and sensitivity to cyanide and hydrogen peroxide.

2. Mn SOD in the mitochondrial matrix and in prokaryotes and is insensitive to cyanide.

3. Fe SOD, usually found in prokaryotes and in the chloroplasts of some plants. It is not sensitive to cyanide but is inhibited by hydrogen peroxide.

4. Al SOD has recently reported [7].

Catalase: H_2O_2 is also metabolized by catalase (CAD), a heme protein with an extremely high turnover rate

$$2\,H_2O_2 \rightarrow 2\,H_2O + O_2$$

SOD protects from senescence, aging, ischemic tissue damage, lipid peroxidation, protein denaturation, and radiation damage.

Glutathione peroxidase: Glutathione carries out the reduction of H_2O_2 which is enzymatic reaction catalyzed by GPx, found in vacuole, cystol, and extracellular space. The enzyme has substrate specificity. Peroxidases are involved in (1) biotic and abiotic stresses, (2) lignin and suberin synthesis, and (3) disease and pathogen response [8].

$$2\,GSH + H_2O_2 \rightarrow GSSG + 2\,H_2O$$

Consequence of H_2O_2 accumulation in glucose-6-phosphate dehydrogenase deficiency due to malarial drug primaquine results in hemolytic anemia due to oxidative stress.

Glutathione reductase: Glutathione keeps cysteine thiol groups in the reduced state. If two thiol groups become oxidized, they can be reduced nonenzymatically by glutathione. GSSG is reduced by NADPH-dependent enzyme glutathione reductase.

$$GSSG + NADPH + H \rightarrow 2GSH + NADP^+$$

Glutathione-S-transferases: Through the action of this widely distributed enzyme, glutathione participates in detoxification of xenobiotics or foreign organic compounds.

Glutathione: Glutathione is a tripeptide that is present in high concentrations in most eukaryotic cells and reacts with free radicals. It directly quenches lipid peroxides. Vitamin C and glutathione work interactively [9].

2.2 Nonenzymatic antioxidants

These are biological molecules that can act as antioxidants by either quenching a free radical directly or indirectly by promoting a process responsible for radical scavenging indirectly [10].

a. **Selenium:** Selenium is a mineral and a component of antioxidant enzymes. Rice and wheat are the major dietary sources of selenium. The amount of selenium in soil, which varies by region, determines the amount of selenium in the foods grown in that soil. Animals that eat grains or plants grown in selenium-rich soil have higher levels of selenium in their muscle. Brazil nuts also contain large quantities of selenium.

b. **Transferrin:** Transferrin is a major iron transporting protein in the body. It is normally 20–30% loaded.

c. **Lactoferrin:** Lactoferrin is a milk protein similar to transferrin that helps in iron binding.

d. **Ceruloplasmin:** Ceruloplasmin catalyzes the oxidation of Fe^{++} to Fe^{+++}, while oxygen is reduced to water.

e. **Vitamin A:** Vitamin A is found in three main forms: retinol (vitamin A1), 3,4-didehydroretinol (vitamin A2), and 3-hydroxyretinol (vitamin A3). Foods rich in vitamin A include liver, sweet potatoes, carrots, milk, egg yolks, and mozzarella cheese.

f. **Vitamin C (ascorbic acid):** In the aqueous phase, ascorbic acid may reduce reactive oxygen metabolites directly, with the concurrent formation of dehydroascorbate and/or indirectly by the regeneration of tocopherol from the tocopherol radical [11]. Vitamin C can be found in high abundance in many fruits and vegetables and is also found in cereals, beef, poultry, and fish.

g. **Vitamin E:** Vitamin E, also known as alpha-tocopherol, is found in almonds and oils, including wheat germ, safflower, corn, and soybean oils, and is also found in mangoes, nuts, broccoli, and other foods [12]. It reacts with reactive oxygen metabolites, yielding lipid hydroperoxide, which can be removed by the activity of the phospholipase-GSPHx system.

h. **β-Carotene:** β-Carotene is a lipid-soluble precursor of vitamin A. It functions synergistically with tocopherol to prevent lipid peroxidation.

i. **Ubiquinol-10:** It is a reduced form of coenzyme Q10, present in lipoprotein at relatively low concentrations. It probably regenerates tocopherol from the tocopheroxyl radical and increases its antioxidant efficiency.

2.3 Plant-derived antioxidants

To protect the cells and organ systems of the body against ROS, humans have evolved a highly sophisticated and complex antioxidant protection system. It involves a variety of components, both endogenous in origin, that function interactively and synergistically to neutralize free radicals [13].

These components include:

Nutrient-derived antioxidants like ascorbic acid, tocopherols and carotenoids, and other low-molecular weight compounds such as GSH and lipoic acid.

Antioxidant enzymes, for example, SOD, GSHPx and GSH reductase, which catalyze free radical quenching reactions.

Metal-binding proteins such as ferritin, lactoferrin, albumin, and ceruloplasmin that sequester free iron and copper ions as these ions are capable of catalyzing oxidative reactions.

Numerous other antioxidant phytonutrients present in a wide variety of plant foods.

3. Antioxidant operation and mechanisms

The word antioxidant is used in a general sense to refer to any type of chemical agent which inhibits attack by oxygen or ozone [14]. As applied to vegetable oils, antioxidants are compounds which interrupt the oxidation process by preferentially reacting with the fat radical to form a stable radical which does not quickly react with oxygen [15]. Antioxidants function either by inhibiting the formation of free alkyl radicals in the initiation step or by interrupting the propagation of the free radical chain. In truncating the propagation step, the antioxidants function as hydrogen donors. Generally, the most popular antioxidants are hydroxyphenol compounds with various ring substitutions. The antioxidant radical is stabilized with its local electrons delocalized; hence antioxidant free radicals do not readily initiate other free radicals. They rather even react with lipid free radicals to form stable and complex compounds. In investigating phenolic antioxidants, it is found that their antioxidative capabilities bear a relationship to the number of phenol groups occupying 1,2 or 1,4 positions in an aromatic ring as well as to the volume and electronic characteristics of the ring substituents present [16]. In elucidating the mechanism of oxidative inhibition, it is generally established that antioxidants function as oxygen interceptors in the oxidative process thereby breaking the chain reaction that perpetuates the process [17]. The general scheme is presented below:

$$R + AH \rightarrow RH + A$$

$$RO + AH \rightarrow ROH + A$$

$$ROO + AH \rightarrow ROOH + A$$

$$R + A \rightarrow RA$$

$$RO + A \rightarrow ROA$$

$$ROO + A \rightarrow ROOA$$

$$\text{Antioxidant} + O_2 \rightarrow \text{Oxidized antioxidant}$$

Certain metallic ions such as copper and iron act as prooxidants, catalyzing the oxidation process. Such metal ions can be sequestered or chelated by certain organic acids. They effectively contribute to lower transition metal activity. Examples of such compounds are citric acid, phosphoric acid, and some of their derivatives.

4. Estimation of antioxidants

4.1 Conjugated diene assay

This method allows dynamic quantification of conjugated dienes as a result of initial PUFA (polyunsaturated fatty acids) oxidation by measuring UV absorbance at 234 nm. The principle of this assay is that during linoleic acid oxidation, the double bonds are converted into conjugated double bonds, which are characterized by a strong UV absorption at 234 nm. The activity is expressed in terms of inhibitory concentration (IC_{50}) [17, 18, 19].

4.2 DPPH method (1,1 diphenyl-2-picrylhydrazyl)

This most widely reported DPPH assay method is based on the reduction of methanolic solution of colored free radical DPPH by free radical scavenger. The procedure involves measurement of decrease in absorbance of DPPH at its absorption maxima of 516 nm, which is proportional to concentration of free radical scavenger added to DPPH reagent solution. The activity is expressed as effective concentration EC_{50} [20].

4.3 Superoxide radical scavenging activity

In vitro superoxide radical scavenging activity is measured by riboflavin/light/NBT (Nitro blue tetrazolium) reduction. NBT method is based on generation of superoxide radical by auto-oxidation of riboflavin in presence of light. The superoxide radical reduces NBT to a blue-colored formazan that can be measured at 560 nm. The capacity of extracts to inhibit the color to 50% is measured in terms of EC_{50}. Antioxidant activity of *Ailanthus*, flavonoids, and triphala has been reported in terms of superoxide radical scavenging activity. The superoxide radical can also be detected by oxidation of hydroxylamine, yielding nitrite which is measured colorimetric reaction [21, 22].

4.4 Hydroxyl radical scavenging activity

This method involves in the in vitro generation of hydroxyl radicals using Fe^{3+}/ascorbate/EDTA/H_2O_2 system using Fenton reaction. Scavenging of this hydroxyl radical in presence of antioxidant is measured. In one of the methods, the hydroxyl radicals formed by the oxidation is made to react with DMSO (dimethyl sulphoxide) to yield formaldehyde. Formaldehyde formed produces the intense yellow color with Nash reagent (2 M ammonium acetate with 0.05 M acetic acid and 0.02 M acetyl acetone in distilled water). The intensity of yellow color formed by that reaction is measured at 412 nm spectrophotometrically against reagent blank. The activity is expressed as % hydroxyl radical scavengering [21].

4.5 Nitric oxide radical inhibition activity

Nitric oxide, because of its unpaired electron, is classified as a free radical and displays important reactivities with certain types of proteins and other free radicals. In vitro inhibition of nitric oxide radical is also a measure of antioxidant activity. This method is based on the inhibition of nitric oxide radical generated from sodium nitroprusside in buffer saline and measured by Griess reagent. In presence of scavengers, the absorbance of the chromophore is evaluated at 546 nm. The activity is expressed as % reduction of nitric oxide [21].

4.6 Reducing power method

This method is based on the principle of increase in the absorbance of the reaction mixture, which indicates increase in the antioxidant activity. In this method, antioxidant compound forms a colored complex with potassium ferricyanide, trichloroacetic acid, and ferric chloride, which is measured at 700 nm. Increase in absorbance of the reaction mixture indicates the reducing power of the samples [23].

4.7 Phosphomolybdenum method

A spectroscopic method for the quantitative determination of antioxidant capacity, through the formation of phosphomolybdenum complex. The assay is based on the reduction of Mo (VI) to Mo (V) by the sample and subsequent formation of a green phosphate Mo (V) complex at acidic pH [24].

4.8 Peroxynitrite radical scavenging activity

Peroxynitrite is now recognized by researchers as the culprit in many toxic reactions. Hence, an in vitro method for scavenging of peroxy radical has been developed to measure antioxidant activity. The scavenging activity is measured by monitoring the oxidation of dihydrorhodamine on a microplate fluorescence spectrophotometer at 485 nm [25].

4.9 ABTS (2,2-azino-bis(3-ethyl benzothiazoline-6-sulfonic acid) diammonium salt) method

This is a measure of antioxidant activity. It also permits to distinguish between additive and synergistic effects. The assay is based on interaction between antioxidant and ABTS+ radical cation which has a characteristic color showing maxima at 645, 734 and 815 nm [24–26].

4.10 DMPD (N,N-dimethyl-p-phenylenediamine dihydrochloride) method

This assay is based on the reduction of buffered solution of colored DMPD in acetate buffer and ferric chloride. The procedure involves measurement of decrease in absorbance of DMPD in presence of scavengers at its absorption maxima of 505 nm. The activity was expressed as percentage reduction of DMPD [24–27].

4.11 Oxygen radical absorbance capacity (ORAC)

ORAC is an exciting and revolutionary new test tube analysis that can be utilized to test "antioxidant power" of foods and other chemical substances. It calculates the ability of a product or chemical to protect against potentially damaging free radicals. This analytical procedure measures the ability of a substance to act as an antioxidant. The test is performed using Trolox (a water-soluble analog of vitamin E) as a standard to determine the Trolox equivalent (TE). The ORAC value is then calculated from the Trolox equivalent and expressed as ORAC units or value. From this assay it shows the higher the ORAC value, the greater the "antioxidant power." In automated ORAC assay B-phycoerythrin (b-PE) was used as a target free radical damage, AAPH as a peroxy radical generator and Trolox as a standard control. After addition of AAPH to the test solution, the fluorescence is recorded, and the antioxidant activity is expressed as Trolox equivalent [28].

4.12 β-Carotene linoleate model

This is one of the rapid methods to screen antioxidants, which is mainly based on the principle that linoleic acid, which is an unsaturated fatty acid, gets oxidized by "reactive oxygen species" (ROS) produced by oxygenated water. The products formed will initiate the β-carotene oxidation, which will lead to discoloration. Antioxidants decrease the extent of discoloration, which is measured at 434 nm, and the activity is measured [24].

4.13 TRAP method

This method is defined as total radical trapping antioxidant parameter. The fluorescence of R-phycoerythrin is quenched by ABAP (2,2′-azobis(2-amidinopropane) hydrochloride) as a radical generator. The antioxidative potential is evaluated by measuring the delay in decoloration [29].

4.14 Cytochrome c test

Superoxide anions were assayed spectrophotometrically by a cytochrome reduction method described by McCord [6]. Xanthine oxidase converts xanthine to uric acid and yields superoxide anions which directly reduce ferricytochrome c to ferrocytochrome c, having an absorbance change at 550 nm. [30].

4.15 Erythrocyte ghost system

This method involves isolation of erythrocyte ghost cells and the induction of lipid peroxidation using them and the induction of tetra-butyl hydroxy peroxide (t-BHP). Thiobarbituric acid reactive substance (TBARS) produced during the reaction is measured at 535 nm [31].

4.16 Microsomal lipid peroxidation or thiobarbituric acid (TBA) assay

TBA test involves isolation of microsomes from rat liver and induction of lipid peroxides with ferric ions leading to the production of small amount of malondialdehyde (MDA). TBA reacts with MDA to form a pink chromogen, which can be detected spectrophotometrically at 532 nm [32].

5. The potential role of antioxidants in disease

5.1 Oxidative stress and diseases

5.1.1 Nephrotic syndrome

The nephrotic syndrome (NS) is defined by heavy proteinuria (urine total protein excretion greater than 3.5 g/d or total protein-creatinine ratio greater than 3.5 g/g) due to abnormal increase of glomerular permeability and following hypoalbuminemia, hyperlipidemia, and edema. Peroxidation of lipid membranes raises the concentration of their by-product MDA and the consequent lowering of antioxidants as a result of consumption [33]. The combined therapy of antioxidants, minerals with B complex vitamins for treatment of imbalance oxidant/antioxidant status, hyperhomocyst(e) inemia, and deficiency of copper and zinc in nephrotic syndrome patients.

5.1.2 Oxidative stress and neurodegenerative diseases

The brain is exposed throughout life to OS, and certain diseases of the brain and nervous system are thought to involve free radical processes and oxidative damage, either as a primary cause or as a consequence of disease progression.

1. **Alzheimer's disease:** Alzheimer's disease (AD) is a progressive neuropsychiatric disorder of unknown etiology. It is characterized by neuronal degeneration and cognitive deterioration, especially in the elderly [34]. OS has been implicated in the pathogenesis of AD [35] by the finding of several characteristics, such as enhanced lipid peroxidation, in specific areas of the brain in postmortem studies [36]. Several investigators detected an increase in the activity of catalase, superoxide dismutase, glutathione peroxidase, and glutathione reductase in the hippocampus and amygdale.

2. **Cognitive dysfunction in the elderly:** Cognitive impairment is a common problem in the over 65-year age group, progressing to its most devastating form of clinical dementia, usually Alzheimer's dementia, in about 5% of this population [37]. Goodwin noted a correlation between memory function and vitamin C in the blood of healthy volunteers aged 60 or over [38]. Accordingly, Perry found a positive association of memory performance with β-carotene and vitamin C levels in plasma measured twice [39].

3. **Parkinson's disease:** Data from postmortem studies of brains from patients with Parkinson's disease (PD) suggest that OS plays an important role in neural degeneration of the pigmented dopaminergic neurons in the substantia nigra pars compacta (SNpc) [40]. One of the suggested causes of OS in the SNpc is the production of ROS during the normal metabolism of dopamine. In the human SNpc, the oxidation products of dopamine may polymerize to form neuromelanin, which may also be toxic [41, 42]. According to postmortem studies, the SNpc of PD patients shows a significant (60%) reduction in GSH and a moderate (29%) increase in oxidized glutathione (GSSG) levels [43, 44].

4. **Huntington's disease:** Huntington's disease is an autosomal neuronal disorder characterized as a movement disorder caused by repetition of a CAG trinucleotide sequences encoding for a polyglutamine tract at the N terminus of the gene encoding a protein named huntingtin [45]. Several postmortem studies showed increased iron levels in the striatum of patients with Huntington's disease [46].

5. **Amyotrophic lateral sclerosis (ALS):** ALS is characterized by a selective and progressive degeneration of the lower motor neurons in the spinal cord and the upper motor neurons in the cerebral cortex, usually beginning in midlife. OS may be involved in all types of ALS [47]. Levels of vitamin E and malondialdehyde (MDA), as a measure of lipid oxidation, increased over time in mutant CuZnSOD mice, as compared to controls [48].

6. **Schizophrenia and tardive dyskinesia:** The presence of excess levels of ROS has been described for both schizophrenia and neuroleptic-induced tardive dyskinesia [49]. The contribution of oxidative injury to the pathophysiology of schizophrenia is indicated by the increase in lipid peroxidation products in the plasma and CSF and the altered levels of both enzymatic and nonenzymatic antioxidants in chronic naive first-episode patients [50, 51].

7. **Chemically induced neurological disorders:** Several neurotoxic chemicals have been shown to elevate the cerebral rate of ROS production in experimental animals. These include methylmercuric chloride, cadmium, toluene, and other organic solvents [52, 53]. All of these agents are also capable of increasing intracellular levels of calcium ions [54].

8. **Brain aging:** Aging in mammalian species appears to be the result of normal developmental and metabolic processes responsible for graying of the hair, decreases in the rate of wound healing, and increases in susceptibility to disease and death. Studies have found evidence of oxidative damage to macromolecules (DNA, lipids, and proteins) especially in brains from elderly subjects, supporting the hypothesis that oxidative injury might directly cause the aging process [55–57].

5.1.3 Diabetes mellitus

Diabetes in humans is a disease associated with increased oxidative stress. The cause of this is not yet fully understood but is thought to include mitochondrial dysfunction, direct enzyme inhibition by hyperglycemia, auto-oxidation of glucose, and activation of NADPH oxidase. The oxidative stress manifests itself as elevated concentrations of lipid peroxidation products, erythrocyte fragility, and decreases in the antioxidant enzyme systems (CAT, GSH-PX, and SOD) [58–61].

5.1.4 Asthma

Feline asthma closely parallels human asthma, which is known to be associated with oxidative stress. Such cells generate ROS, which are involved in the pathophysiology of asthma [62, 63].

5.1.5 Atherosclerosis

It has been known that LDL can be oxidized by many kinds of oxidants by different mechanisms and pathways. Myeloperoxidase (MPO) secreted from phagocytes has been implicated in the pathogenesis of atherosclerosis. Reactive nitrogen species are another species, which may contribute in atherosclerosis. Nitric oxide (NO) is not a strong oxidant in itself, but it reacts rapidly with O_2 to give peroxynitrite, which oxidizes LDL to an atherogenic form [64].

5.1.6 Heart failure

Accumulating evidence suggests that reactive oxygen species (ROS) play an important role in the development and progression of heart failure, regardless of the etiology.

5.1.7 Hemorrhagic shock

Acute hemorrhagic shock causes decreases in the cardiac function and contractility and is associated with an increase in oxygen free radical (OFR) producing activity of PMN leukocytes [65].

5.1.8 Ischemia–reperfusion

Reactive oxygen-derived radicals and metabolites are known to play important roles in the pathogenesis of ischemia/reperfusion and anoxia/reoxygenation injury. Free radicals are induced by the reperfusion blood flow in addition the lack of oxygen (O_2) supply to the ischemic cell.

5.1.9 Lung disease

The large endothelial surface is constantly exposed to many atmospheric pollutants including tobacco smoke, fuel emissions, ozone, and nitrogen dioxide, and given the natural oxidizing nature of the atmosphere (e.g., 21% O_2), the lung is always at risk of oxidative injury [66].

5.1.10 Aging

The free radical theory of aging includes phenomenological measurements of age-associated oxidative stress, interspecies comparisons, dietary restriction, the manipulation of metabolic activity and oxygen tension, treatment with dietary and pharmacological antioxidants, in vitro senescence, classical and population genetics, molecular genetics, transgenic organisms, the study of human diseases of aging, epidemiological studies, and the ongoing elucidation of the role of active oxygen in biology [67].

5.1.11 Free radicals and cancer

One type of endogenous damage is that arising from intermediates of oxygen (dioxygen)-reduction oxygen free radicals, which attacks not only the bases but also the deoxyribosyl backbone of DNA. OFR are also known to attack other cellular components such as lipids, leaving behind reactive species that in turn can couple to DNA bases [68].

5.1.12 Inflammation

During phagocytosis, cells consume increased amount of oxygen, a process termed the respiratory burst. Activation results in increased NADPH production via the hexose monophosphate shunt, and the generation of O_2, H_2O_2, OH and hypochlorous acid (HOCI), hypoxanthine concentration, xanthine oxidase activity, and ROS production are increased in rheumatoid arthritis [69].

5.1.13 Ocular disease

Oxidative stress is implicated in age-related macular degeneration and cataracts by altering various cell types in the eye either photochemically or nonphotochemically [70]. Under the action of free radicals, the crystalline proteins in the lens can cross-link and aggregate, leading to the formation of cataract [71, 72].

5.1.14 Fetus

Oxidative stress is involved in many mechanisms in the development of fetal growth restriction and preeclampsia in prenatal medicine. Some reports indicate that blood levels of lipid peroxidation products (F2-isoprostanes, MDA) are elevated in preeclamptic pregnancy and intra-uterine growth retardation, and it has been suggested that ROS/RNS play a role in the etiology of these diseases [63, 73]. In pregnancies complicated by preeclampsia, increased expression of NADPH oxidase 1 and 5 isoforms which are the major enzymatic sources of superoxide in the placenta is seen [74].

S. no	Plant name	Family	Part used	Chemical constituents responsible for antioxidant activity	Reference(s)
1	*Amaranthus paniculatus*	Amaranthaceae	Leaf	Carotenoids, ascorbic acid, flavonoids, and phenolic acids	[75]
2	*Amaranthus gangeticus*	Amaranthaceae	Leaf	Carotenoids, ascorbic acid, flavonoids, and phenolic acids	[75]
3	*Amaranthus blitum*	Amaranthaceae	Leaf	Carotenoids, ascorbic acid, flavonoids, and phenolic acids	[75]
4	*Amaranthus spinosus*	Amaranthaceae	Leaf	Carotenoids, ascorbic acid, flavonoids, and phenolic acids	[75]
5	*Amaranthus viridis*	Amaranthaceae	Leaf	Carotenoids, ascorbic acid, flavonoids, and phenolic acids	[75]
6	*Coriandrum sativum*	Umbelliferae	Leaf, fruit	S-(+)-linalool, monoterpenes, hydrocarbons, namely, α-pinene, limonene, γ-terpinene, p-cymene, borneol, citronellol, camphor, geraniol, and geraniol acetate, heterocyclic components like pyrazine, pyridine, thiazole, furan and tetrahydrofuran derivatives, isocoumarins, coriandrin, dihydrocoriandrin, coriandrons A–E, flavonoids, pthalides, neochidilide, digustilide phenolic acids, and sterols	[76]
7	*Emblica officinalis*	Umbelliferae	Fruit, leaves	Vitamins, ascorbic acid, and phenolics are known as hydrophilic antioxidants, while carotenoids are known as lipophilic antioxidants	[76]
8	*Digera muricata* (L.)	Amaranthaceae	Leaf	Phenols, flavonoids, glycosides, tannins and terpenoids, and minimum for saponins	[76]
9	*Chenopodium album* L.	Amaranthaceae	Leaf	Alkaloids, apocarotenoids, flavonoids, phytoecdysteroids xyloside, limonene (23.2%), α-terpinyl acetate (13.7%), α-terpinene (12.3%), and cis-ascaridole (12.2%)	[77]

S. no	Plant name	Family	Part used	Chemical constituents responsible for antioxidant activity	Reference(s)
10	*Basella alba Linn*	Basellaceae	Leaf	Proteins, fat, vitamin A, vitamin C, vitamin E, vitamin K, vitamin B9 (folic acid), riboflavin, niacin, thiamine, and minerals such as calcium, magnesium, iron	[78]
11	*Basella rubra*	Basellaceae	Leaf	Calcium, iron, vitamins A, B, and C, saponins A, B, C, and D, oleanane-type triterpene oligoglycosides, spinacoside C, and momordins IIb and IIc, β-carotene, small amounts of α-carotenes, 4-coumaroyl, and feruloyl derivatives	[79–81]
12	*Physalis philadelphica*	Solanaeae	Leaf, fruit	2,3-Dihydro-3beta-methoxyixocarpalactone A, 2,3-dihydro-3beta-methoxyixocarpalactone B, 2,3-dihydroixocarpalactone B	[82]
13	*Rumex vesicarius*	Polygonaceae	Leaf	Minerals, protein and ascorbic acid, oxalic acid, tocopherol and lipids. Ca, Cu, Fe, Mg, K, Na, Zn, lipids, ascorbic acid, tocopherol	[83]
14	*Paederia foetida*	Rubiaceae	Leaves	B-Sitosterol, leupiol, methyl mercaptan, crystalline keto alcohol, paederolone, paederone, and hetasitosterol	[84]
15	*Solanum nigrum Linn*	Solanaceae	Leaf	Acetic acid, tartaric acid, malic acid and citric acid, solanine, alpha, beta gamma chaconines, and alpha, beta gamma solanines, solanine, beta-2-solamargine, solamargine, and degalactotigonin. Five non-saponins including p-hydroxybenzoic acid and 3-methoxy-4-hydroxybenzoic acid	[85]
16	*Trigonella foenum-gracecum Linn*	Leguminosae	Leaf	Amino acid, fatty acid, vitamins, saponins, folic acid, disogenin, gitogenin, neogitogenin, homorientinsaponaretin, neogigogenin, and trigogenin, 4,5[delta]-cadinene (27.6%), [á]-cadinol, palmitic acid, linoleic acid, oleic acid and stearic acid, hexanal, 2-methyl-2-butenal, 3-octen-2-one, flavonoids, polysaccharides, saponins, polysaccharides, trigonelline, choline, quercetin, galactomannan, polysaccharides	[86]
17	*Brassica oleracea* Capitata	Brassicaceae	Leaf	Glucosinolates and their derived products, flavonoids, and other phenolics, quercetin 3-O-sophoroside-7-O-glucoside, 3-p-coumaroylquinic acid, kaempferol-3-O-sophoroside-7-O-glucoside, kaempferol 3-O-(caffeoyl)-sophoroside-7-O-glucoside, sinapoyl glucoside acid, kaempferol 3-O-(sinapoyl)-sophoroside-7-O-glucoside, sinapic acid, 3 isomeric forms of 1,2-disinapoylgentiobiose, kaempferol 3-O-sophoroside-7-O-glucoside	[87]
18	*Moringa pterygosperma Gaertn*	Moringaceae	Leaf	4-(4'-*O*-Acetyl-α-L-rhamnopyranosyloxy) benzyl isothiocyanate, 4-(α-L-rhamnopyranosyloxy)benzyl isothiocyanate, niazimicin, pterygospermin, benzyl isothiocyanate, and 4-(α-L-rhamnopyranosyloxy)benzyl glucosinolate, carotenoids (including β-carotene or pro-vitamin A)	[88]

S. no	Plant name	Family	Part used	Chemical constituents responsible for antioxidant activity	Reference(s)
19	*Hibiscus cannabinus L*	Malvaceae	Leaf	Tannins, saponins, polyphenolics, alkaloids, lignans, essential oils, and steroids	[89]
20	*Sesbania grandiflora L*	*Fabaceae*	Leaf	Galactomannans, linoleic acid, β-sitosterol, and carbohydrates. Vitamin C, and calcium, iodine, pectin, saponins, aliphatic alcohol, leucocyanidin and cyanidin, oleanolic acid and its methyl ester and kaempferol-3-rutinoside, tannins and gum, sesbanimide	[90–92]
21	*Portulaca oleracea L*	Portulacaceae	Leaf	Omega-3 fatty acids, gallotannins, kaempferol, quercetin, apigenin, α-tocopherols, ascorbic acid and glutathione, free oxalic acids, β-carotene, omega-3 fatty acids, coumarins, flavonoids, monoterpene glycoside, and anthraquinone glycosides	[93–95]
22	*Murraya koenigii L*	Rutaceae	Leaf	Alkaloid, volatile oil, glycozoline, xanthotoxin, and sesquiterpine	[96–100]
23	*Celosia argentea*	Amaranthaceae	Leaf	Alkaloids, glycosides, flavonoids, saponins, tannins, carbohydrate and essential oils, steroids, carotenoids, and anthocyanins	[101]
24	*Boerhavia diffusa*	Nyctaginaceae	Leaf	Alkaloids, punarnavine, rotenoids (boeravinones A–F), amino acids, lignans (liriodendrons), ß-sitosterols and tetracosanoic, esacosanoic, stearic, and ursolic acids. Rotenoids known as boeravinones, punarnavoside, a phenolic glycoside, 11,12 C-methyl flavone liriodendrin and syringaresinolmono-β-D-glycoside, fatty acids and allantoin boerhavin and boerhavic acid, aegeline, aegelinine, rutin, sterol, tannins, flavonoids, quercetin, volatile oils, ß-sitosterols	[102–109]
25	*Eclipta alba*	Asteraceae	Leaf	Coumestans, alkaloids, flavonoids, glycosides, polyacetylenes, triterpenoids, and thiophenes. Phytosterol, P-amyrin, luteolin-7-glucoside, P-glucoside of phytosterol, a glucoside of a triterpenic acid and wedelolactone. Cystine, glutamic acid, phenylalanine, tyrosine and methionine, nicotine, and nicotinic acid	[110]
26	*Centella asiatica*	Apiaceae	Leaf	Asiaticoside carotene, ascorbic acid, phenols, madecassic acid	[111]
27	*Phyllanthus amarus*	Euphorbiaceae	Leaf	Alkaloids, astragalin, brevifolin, carboxylic acids, corilagin, cymene, ellagic acid, ellagitannins, gallocatechins, *geraniin*, hypophyllanthin, phyllanthin, lignans, lintetralins, lupeols, methyl salicylate, phyllanthine, phyllanthenol, phyllochrysine, phyltetralin, repandusinic acids, quercetin, quercetol, quercitrin, rutin, saponins, triacontanal, and tricontanol	[112]
28	*Hibiscus sabdariffa*	Malvaceae	Leaf	Ascorbic acid (vitamin C) and tocopherol (vitamin E), flavonoids, polyphenols	[83]

S. no	Plant name	Family	Part used	Chemical constituents responsible for antioxidant activity	Reference(s)
29	*Curcuma longa*	Zingiberaceae	Leaf	Ascorbic-acid rhizome,beta-carotene rhizome, caffeic-acid rhizome, curcumin rhizome, eugenol essential oil, p-coumaric-acid rhizome, protocatechuic acid leaf, syringic-acid leaf, vanillic acid in leaf, camphene, eugenol, curcumin	[113]
30	*Ocimum sanctum*	Labiatae	Leaf	Volatile oil, terpenoids, eugenol, thymol, estragole	[114]
31	*Basella alba*	Basellaceae	Leaf	High in vitamin A, vitamin C, Ca, Iron, phosphorus, vitamin B9 (folic acid), calcium, magnesium, flavonoids, polyphenols	[115]
32	*Mentha arvensis*	Labiatae	Leaf	Flavonoids, acacetin, chrysoeriol, diosmin, eriocitrin, hesperidin, luteolin, esperidoside, menthoside, methyl rosmarinate, rutin, tilianine, narirutin, and nodifloretin. Phenolic acids such as caffeic acid, lithospermic acid, rosmarinic acid, protocatechuic acid, protocatechuic aldehyde, phytosterols, β-sitosterol, and daucosterol; the anthraquinones aloe-emodin,emodin, chrysophanol, and tannins	
33	*Alternanthera sessilis*	Amaranthaceae	Leaf	Carotenoids, triterpene, saponins, flavonoids, steroids, stigmasterol, β-sitosterol, glycosides, protein and amino acids, campesterol, lupeol	[116]
34	*Rumex acetosa*	Polygonaceae	Leaf	Oxalates, including calcium oxalate and tannins; anthracene derivatives, emodin, rhein, quinoids, and flavonoids	[117]
35	*Spinacia oleracea*	Amaranthaceae	Leaf	Vitamin A (especially high in lutein), vitamin C, vitamin E, vitamin K, magnesium, manganese, folate, betaine, iron, vitamin B2, calcium, potassium, vitamin B6, folic acid, copper, protein, phosphorus, zinc, niacin, selenium, and omega-3 fatty acids. Recently, opioid peptides called rubiscolins have also been found in spinach. It is a source of folic acid	[117]
36	*Trianthema portulacastrum*	Aizoaceae		Tetraterpenoid 1 (trianthenol) flavonoid, 5,7-dihydroxy-6,8-dimethylchromone (leptorumol) Isoamericanin A	[118–121]
37	*Hibiscus sabdariffa*	Malvaceae	Leaf	Alkaloids, saponins, tannins, anthraquinones, cardiac glycosides, flavonoids and phlobatannins	[122]

6. Conclusion

The most important free radical in biological systems is radical derivatives of oxygen with the increasing acceptance of free radical as common place and important biochemical intermediate. Antioxidants are believed to play a very important role in the body defense system against reactive oxygen species (ROS), which are the harmful by-products generated during normal cell aerobic respiration. The imbalance between ROS and antioxidant defense system increases the oxidation burden and leads to the damage of macromolecules such as carbohydrates or proteins, such processes of various diseases. To protect the cells and organ systems of the body against reactive oxygen species, humans have evolved a highly sophisticated and complex antioxidant protection system. Plants having vitamins (C, E, carotenoids, etc.), flavonoids (flavones, isoflavones, flavanones, anthocyanins, and catechins), polyphenols (ellagic acid, gallic acid, and tannins) possess remarkable antioxidant activity. Antioxidant activity is neither restricted to a particular part of plant nor the specific families. Current review reveals the different potential application of antioxidant/free radical manipulations in prevention or control of diseases. All plants discussed in this review exhibited significant, clinical, and pharmacological activity with fewer side effects.

Author details

Arun Rasheed* and Rinshana Fathima Abdul Azeez
Department of Pharmaceutical Chemistry, Al Shifa College of Pharmacy, Malappuram, Kerala, India

*Address all correspondence to: arunrasheed@rediffmail.com

References

[1] Liu F, Ooi VEC, Chang ST. Free radical scavenging activity of mushroom polysaccharide extracts. Life Sciences. 1997;**60**:763-771

[2] Droge W. Free radicals in the physiological control of cell function. Physiological Reviews. 2002;**82**:47-95

[3] Sies H. Antioxidants in Disease, Mechanisms and Therapy. New York: Academic Press; 1996

[4] Xianquan S, Shi J, Kakuda Y, Yueming J. Stability of lycopene during food processing and storage. Journal of Medicinal Food. 2005;**8**(4):413-422

[5] Hurrell R. Influence of vegetable protein sources on trace element and mineral bioavailability. Journal of Nutrition. 2003;**133**(9):2973-2977

[6] Fridovich I. Superoxide radical and superoxide dismutase. Biochemical Society Transactions. 1973;**1**:48

[7] Cadmak L, Horst WJ. Effect of aluminium on lipid peroxidation, superoxide dismutase, catalase and peroxidase activities in root tips of Soybean. Plant Physiology. 1991;**83**:463-468

[8] Chen S, Schopfer P. Hydroxyl radical production in physiological reactions: A novel function of peroxidase. European Journal of Biochemistry. 1999;**260**:726-735

[9] Kumar G, Sharmila Banu G, Vanitha Pappa P, Sundararajan M, Rajasekara PM. Hepatoprotective activity of *Trianthema portulacastrum* L against paracetamol and thioacetamide intoxication in albino rats. Journal of Ethnopharmacology. 2004;**92**:37-40

[10] Irshad M, Chaudhary PS. Oxidant-antioxidant system: Role and significant in human body. Indian Journal of Experimental Biology. 2002;**40**:1233-1239

[11] Packer JE, Slater TF, Wilson RS. Direct observation of free radical interaction between vitamin E and Vitamin C. Nature. 1979;**278**:737-738

[12] Herrera E, Barbas C. Vitamin E: Action, metabolism and perspectives. Journal of Physiology and Biochemistry. 2001;**57**(2):43-56

[13] Mark P. Antioxidants. Clinical Nutrition Insight. 1998:NUT031, 1/96 Rev 10/98, 1-4

[14] Scott G. Atmospheric Oxidation and Antioxidants. Amsterdam: Elsevier Publishing Company; 1965

[15] Eastman Chemical Company. High Performance Additives. Kingsport, TN, USA: Eastman Chemical Company; 2007. www.eastman.com

[16] Fennema OR. Food Chemistry. 2nd ed. New York: Marcell Dekker, Inc.; 1985. pp. 46-50

[17] Bennion M. Introductory Foods. 10th ed. Upper Saddle River, New Jersey, USA: Prentice-Hall Inc.; 1995

[18] Ashok KJ. Imbalance in antioxidant defence and human diseases: Multiple approach of natural antioxidant therapy. Current Science. 2001;**81**(9):1179-1186

[19] David GB, Erik EA, Rohini S, Alfins. Antioxidant enzyme expression and ROS damage in prostatic intraepithelial neoplasia and cancer. Cancer. 2000;**89**:124-134

[20] Sanchez-Moreno C, Larrauri J, Saura-Calixto F. Free radical scavenging capacity of selected red and white wine. Journal of the Science of Food and Agriculture. 1999;**79**:1301-1304

[21] Babu BH, Shylesh BS, Padikkala J. Antioxidant and hepatoprotective effect of *Alanthus icicifocus*. Fitoterapia. 2001;**72**:272-277

[22] Robak J, Gryglewski RJ. Flavonoids are scavengers of superoxide anions. Biochemical Pharmacology. 1998;**37**:837-841

[23] Jayaprakash GK, Singh RP, Sakariah KK. Antioxidant activity of grape seed extracts on peroxidation models in-vitro. Journal of Agricultural and Food Chemistry. 2001;**55**:1018-1022

[24] Kanner J. Natural antioxidants in grapes and wines. Journal of Agricultural and Food Chemistry. 1994;**42**:64-69

[25] Choi HR. Peroxynitrite scavenging activity of herb extracts. Phytotherapy Research. 2002;**16**:364-367

[26] Simonetti P, Pietta P, Testolin G. Polyphenol content and total antioxidant potential selected Italian wines. Journal of Agricultural and Food Chemistry. 1997;**45**:1152-1155

[27] Vinson JA, Hontz BA. Phenol antioxidant index: Comparative antioxidant effectiveness of red and white wines. Journal of Agricultural and Food Chemistry. 1995;**43**:401-403

[28] Ronald LP. Anti oxidant capacity as influenced by total phenolic & anthocyanin content maturity and variety of *Vaccinium* species. Journal of Agricultural and Food Chemistry. 1998;**46**:2686-2693

[29] Ghiselli A. Fluorescence based method for measuring total plasma antioxidant capability. Free Radical Biology & Medicine. 1995;**18**:29-36

[30] Ho KY, Huang JS, Tsai CC, Lin TC, Lin CC. Antioxidant activity of tannin component from *Vaccinium vitis-idaea*. Journal of Pharmacy and Pharmacology. 1999;**51**:1075-1078

[31] Chiaki S, Naomi. Antioxidative polyphenols isolated from *Theobroma cacao*. Journal of Agricultural and Food Chemistry. 1998;**46**:454-457

[32] Gutteridge JMC, Wilkins S. Copper salt dependent hydroxyl radical formation. Damage to proteins acting as antioxidant. Biochimica et Biophysica Acta. 1986;**754**:38-41

[33] Sanjay K, Bimbadhar R, Bhaskar CK. Indirect quantification of lipid peroxidation in steroid responsive nephrotic syndrome. Archives of Disease in Childhood. 2000;**82**:76-78

[34] Zachwieja J, Bobkawski W, Niklas A. Total antioxidant status in children with nephrotic syndrome. Polski Merkuriusz Lekarski. 2000;**38**(46):216-217

[35] Leszek T, Boleslaw R, Watter HH. Antioxidants: Possible role in kidney protection. Kidney & Blood Pressure Research. 2003;**26**:303-314

[36] Flynn BL, Runho A. Pharmacological management of Alzheimer's disease part II: Antioxidants, antihypertensives and Ergoloid derivatives. The Annals of Pharmacotherapy. 1999;**33**:188-197

[37] Markesbery WR. Oxidative stress hypothesis in Alzheimer's disease. Free Radical Biology and Medicine. 1997;**23**:134-147

[38] Lovell MA, Ehmann WD, Butler SM, Markesberg WR. Elevated thiobarbituric acid reactive substances and antioxidant enzyme activity in the brain in Alzheimer's disease. Neurology. 1995;**45**:1594-1601

[39] Frautschy SA, Baired A, Cole GM. Effects of injected Alzheimer α-amyloid cores in rat brain. Proceedings of the National Academy of Sciences of the United States of America. 1991;**88**:8362-8366

[40] Hoffman A, Grobbee DE, De Jong PTVM, Van den Ouweland A. Determinants of disease and disability in the elderly the Rotterdam Elderly Study. European Journal of Epidemiology. 1991;**7**:403-412

[41] Goodwin JS, Goodwin JM, Garry PJ. Association between nutritional status and cognitive functioning in a healthy elderly population. Journal of the American Medical Association. 1983;**249**:2917-2921

[42] Perry WJ, Perry P, Stahelin HB. The relation between antioxidants and memory performance in the old and very old. Journal of the American Geriatrics Society. 1997;**45**:718-724

[43] Fahn S, Cohen G. The oxidant stress hypothesis in Parkinson's disease. Evidence supporting it. Annals of Neurology. 1992;**32**:804-812

[44] Offen D, Ziv I, Gorodin S, Barzilai A, Malik A, Melamed E. Dopamine induced programmed cell death in mouse thymocytes. Biochimica et Biophysica Acta. 1995;**1268**:171-177

[45] Akaneya Y, Takahasi M, Hatanaka H. Involvement of free radicals in neurotoxicity against rat dopaminergic neurons in culture. Neuroscience Letters. 1995;**193**:53-56

[46] Sian J, Dexter DT, Less AJ. Alteration in glutathione levels in Parkinson's disease and other neurodegenerative disorders affective basal ganglia. Annals of Neurology. 1994;**36**:348-355

[47] Damier P, Hirsch EC, Zhang P, Agid Y, Javoy-Agid F. Glutathione peroxidase, glial cells and Parkinson's disease. Neuroscience. 1993;**52**:1-7

[48] Bartzokis G, Cummings J, Perlman S, Hance DB, Mintz J. Increased basal ganglia iron levels in Huntington's disease. Archives of Neurology. 1999;**56**:569-574

[49] Chen JC, Hurdy DA, Hucharczyk W. MRI of human postmortem brain tissues correlative study between T2 and assays of iron and ferritin in Parkinson and Huntington's disease. American Journal of Neuroscience Research. 1993;**14**:275-281

[50] Gu M, Gash MT, Mann VM, Jany-Agid F, Cooper JM, Schapira AH. Mitochondrial defect in Huntington's disease caudate nucleus. Annals of Neurology. 1996;**39**:385-389

[51] Browne SE, Bowling AC, MacGarrey U. Oxidative damage and metabolic dysfunction in Huntington's disease: Selective vulnerability of the basal ganglia. Annals of Neurology. 1997;**41**:646-653

[52] Oteiza PI, Uchitel OD, Carrasquedo F, Duborovski AL, Roma JC, Fraga CG. Evaluation of antioxidants, protein, and lipid oxidation products in blood from sporadic amyotrophic lateral sclerosis patients. Neurochemical Research. 1997;**22**(4):535-539

[53] Lohr JB, Kuczenski R, Bracha HS, Moir M, Joste DV. Increased indices of free radical activity in the cerebrospinal fluid of patients with tardive dyskinesia. Biological Psychiatry. 1990;**28**:533-539

[54] Reynolds GP. Developments in the drug treatment of schizophrenia. Trends in Pharmacological Sciences. 1992;**13**:116-121

[55] Mahadik SP, Scheffer RE. Oxidative injury and potential use of antioxidants in schizophrenia. Prostaglandins, Leukotrienes and Essential Fatty Acids. 1996;**55**:45-54

[56] Tsai G, Goff DC, Chang RW, Flood J, Baer L, Coyle JT. Markers of glutamatergic neurotransmission and oxidative stress associated with tardive dyskinesia. American Journal of Psychiatry. 1998;**155**(9):1207-1213

[57] Lebel CP, Ali SF, McKee M, Bondy SC. Organometal induced increases in oxygen reactive species: The potential of 2_7_-dichlorofluorescin diacetate as an index of neurotoxic damage. Toxicology and Applied Pharmacology. 1990;**104**:17-24

[58] Mattia CJ, Adams JD, Bondy SC. Free radical induction in the brain and liver by products of toluene catabolism. Biochemical Pharmacology. 1993;**46**:103-110

[59] Bondy SC, Komulainen H. Intracellular calcium as an index of neurotoxic damage. Toxicology. 1988;**49**:35-41

[60] Cutler RG. Human longevity and aging: Possible role of reactive oxygen species. Annals of the New York Academy of Sciences. 1991;**621**:1-28

[61] Harman D. Role of free radicals in aging and disease. Annals of the New York Academy of Sciences. 1992;**673**:126-134

[62] Beal M. Aging, energy and OS in neurodegenerative diseases. Annals of Neurology. 1995;**38**:357-366

[63] Map PI, Grootveld MC, Bike DR. Oxidative stress and rheumatoid arthritis. British Medical Bulletin. 1995;**51**:419-436

[64] Pieper GM, Jordan M, Dondlinger LA, Adams MB, Roza AM. Peroxidative stress in diabetic blood vessels. Diabetes. 1995;**44**:884-889

[65] Collins AR, Raslova K, Somorovska M, Petrovska H, Ondrusova A, Vohnout B, et al. DNA damage in diabetes: Correlation with a clinical marker. Free Radical Biology and Medicine. 1998;**25**:373-377

[66] Duthie SJ. Antioxidant supplementation decreases oxidative DNA damage in human lymphocytes. Cancer Research. 1996;**56**:1291-1295

[67] Smith LJ, Shamsuddin M, Sporn PHS, Denenberg M, Anderson J. Reduced superoxide dismutase in lung cells of patients with asthma. Free Radical Biology & Medicine. 1997;**22**:1301-1307

[68] Gordon RE, Shaked AA, Solano DF. Taurine protects hamster bronchioles from acute NO_2-induced alterations. A histological, ultrastructural and freeze-fracture study. American Journal of Pathology. 1986;**125**:585-600

[69] Niki E. Antioxidants and atherosclerosis. Biochemical Society Transactions. 2004;**32**(1):156-159

[70] Byrne JA, Grieve DJ, Cave AC, Shah AM. Oxidative stress and heart failure. Archives des Maladies du Coeur et des Vaisseaux. 2003;**96**(3):214-221

[71] Asano G, Takashi E, Ishiwata T, Onda M, Yokayama M, Naito Z. Pathogenesis and protection of ischemia and reperfusion injury in myocardium. Journal of Nippon Medical School. 2003;**70**(5):384-392

[72] Beckman KB, Ames BN. The free radical theory of ageing matures. Physiological Reviews. 1998;**78**(2):547-581

[73] Valko M, Izakovic M, Mazur MCJ, Telser J. Role of oxygen radicals in DNA damage and cancer incidence. Molecular and Cellular Biochemistry. 2004;**266**:37-56

[74] Santosa S, Jones PJ. Oxidative stress in ocular disease: Does lutein play a protective role? Canadian Medical Association Journal. 2005;**173**:861-862

[75] Meyer CH, Sekundo W. Nutritional supplementation to prevent cataract formation. Developments in Ophthalmology. 2005;**38**:103-119

[76] Beatty S, Koh HH, Phil M, Henson D, Boulton M. The role of oxidative

stress in the pathogenesis of age-related macular degeneration. Survey of Ophthalmology. 2000;**45**:115-134

[77] Hracsko Z, Orvos H, Novak Z, Pal A, Varga IS. Evaluation of oxidative stress markers in neonates with intra-uterine growth retardation. Redox Report. 2008;**13**:11-16

[78] Biki A, Bozkurt N, Turp A. Role of oxidative stress in intrauterine growth restriction. Gynecologic and Obstetric Investigation. 2007;**64**:187-192

[79] Brazdova K, Krmelova V, Rada K, Starhova H. Anthracene derivatives in *Rumex* species. II. Anthraquinone content in some *Rumex* species. Scientia Pharmaceutica. 1967;**35**:116

[80] Hunter KJ, Fletcher JM. The antioxidant activity and com Leaf position of fresh, frozen, jarred and canned vegetables. Innovative Food Science and Emerging Technologies. 2002;**3**:99-406

[81] Mety S, Mathad P, Rajanna L. Systematic evaluation of free radical scavenging and antioxidative activities in *Digera muricata* (L.) Mart. Asian Journal of Pharmacy and Life Science. 2011;**1**(3)

[82] Adedapo A, Jimoh F, Afolayan A. Comparison of the nutritive value and biological activities of the acetone, methanol and water extracts of the leaves of *Bidens pilosa* and chenopodium album. Acta Poloniae Pharmaceutica. Drug Research. 2011;**68**:83-92

[83] Adhikari R, Naveen Kumar HN, Shruthi SD. A review on medicinal importance of *Basella alba* L. International Journal of Pharmaceutical Science and Drug Research. 2012;**4**(2):110-114

[84] Grubben GJH. PROTA Foundation. Wageningen: Wageningen/Backhuys/ Leiden/CTA; 2004

[85] Penteado MDVC, Minazzi RS, Regina S, Bicuda DAL. Carotinoids and provitamin A. Activity of vegetable leaves consumed in Northern Brazil. Chemical Abstracts. 1987;**107**:609

[86] Glaessgen WE, Metzger JW, Heuer S, Strack D. Betacyanins from fruits of *Basella rubra*. Phytochemistry. 1993;**33**(6):1525-1527

[87] Maldonado E, Pérez-Castorena AL, Garcés C, Martínez M. Philadelphicalactones C and D and other cytotoxic compounds from *Physalis philadelphica*. Steroids. 2011;**76**(7):724-728

[88] Mohamed R, Fernandez J, Pineda M, Aguilar M. Roselle (*Hibiscus sabdariffa*) Seed oil is rich source of Y-tocopherol. Journal of Food Science. 2007;**72**:3

[89] Hossain MM, Ali MS, Saha A,Alimuzzaman M. Antinocicceptive activity of whole plant extracts of *Paederia foetida*. Journal of Pharmaceutical Sciences. 2006;**5**(1):67-69

[90] Atanu FO, Ebiloma UG, Ajayi EI. A review of the pharmacological aspects of *Solanum nigrum* Linn. Biotechnology and Molecular Biology Reviews. 2011;**6**(1):001-007

[91] Toppo FA, Akhand R, Pathak AK. Pharmacological actions and potential uses of *Trigonella foenum-graecum*: A review. Asian Journal of Pharmaceutical and Clinical Research. 2009;**2**:4

[92] Ferreres F, Sousa C, Vrchovská V, Valentão P, Pereira JA, Seabra RM, et al. Chemical composition and antioxidant activity of Tronchuda cabbage internal leaves. European Food Research and Technology. 2006;**222**:88-98

[93] Fuglie LJ. The Miracle Tree: *Moringa oleifera*: Natural Nutrition for the Tropics. Dakar: Church World Service; 1999. pp. 68-172

[94] Moujir L, Seca AML, Silva AMS, López MR, Padilla N, Cavaleiro JAS, Neto CP. Cytotoxic activity of lignans from *Hibiscus cannabinus* Fitoterapia. 2007;**78**(5):385-387

[95] Mendoza VB. Katturai: A Plant of Many Uses. Canopy International; August 1980. pp. 12-13

[96] Devdatta A. Nutritive value of Indian foods. Indian Academy of Sciences. 1954;**398**:297

[97] Anonymous. The Wealth of India. A Dictionary of Indian Raw Materials. New Delhi, India: Council of Scientific and Industrial Research; Vol. 10. 1976

[98] Cai Y, Luo Q, Sun M, Corke H. Antioxidant activity and phenolic compounds of 112 traditional Chinese medicinal plants associated with anticancer. Life Sciences. 2004;**74**:2157-2184

[99] Simopoulos AP. Omega-3 fatty acids and antioxidants in edible wild plants. Biological Research. 2004;**37**:263-277

[100] Radhakrishnan R, Zakaria MN, Islam MW, Chen HB, Kamil M, Chan K, et al. Neuropharmacological actions of *Portulaca oleraceae* L v. sativa (Hawk). Journal of Ethnopharmacology. 2001;**761**:71-176

[101] Atta-ur-Rahman, Zaidi R, Firdous S. NMR Studies on Mahanine, *Fitoterapia*. 1998;**59**(6):494-501

[102] Wong KC, Tie DY. The essential oil of the leaves of *Murraya Koenigii* Spreng. Journal of Essential Oil Research. 1993;**5**(4):371-374

[103] Lal RK, Sharma JR, Naqvi AA, Singh N. Phenotypic and genotypic variability for morphological traits and essential oil components in diverse origin germplasm lines of Curry neem (Murraya koenigii). Journal of Medicinal and Aromatic Plant Sciences. 2001;**23**:392-398

[104] Adebajo AC, Reisch J. Minor furocoumarins of *Murraya koenigii*, Fitoterapia. 2000;**71**(3):334-337

[105] Onayade OA, Adebajo AC. Composition of the Leaf Volatile Oil of *Murraya koenigii* Growing in Nigeria, Journal of Herbs Spices & Medicinal Plants. 2000;**7**(4):59-66

[106] Bhujbal S, Patil K, Patil M. Evaluation of anti pyretic potentials of *Celosia argentea* Linn leaf extract. Planta Indica. 2006;**2**:19-20

[107] Misra AN, Tewari HP. Constituents of roots of *Boerhaavia diffusa*, Phytochemistry. 1971;**10**(12):3318-3319

[108] Lami N, Kadota S, Tezuka Y, Kikuchi T. Constituents of the roots of *Boerhaavia diffusa* Linn. II. Structure and stereochemistry of a new rotenoid boeravinone C2. Journal of Chemical and Pharmaceutical Research. 1990;**38**(6):1558-1562

[109] Lami N, Kadota S, Kikuchi T. Constituents of the roots of *Boerhaavia diffusa* Linn. IV. Isolation and structure determination of boeravinones D, E and F. Chemical and Pharmaceutical Bulletin. 1992;**39**(7):1863-1865

[110] Seth RK, Khanna M, Chaudhary M, Singh S, Sarin JPS. Estimation of punarnavosides, a new antifibrinolytic compound from Boerhaa via diffusa. Indian Drugs. 1986;**23**:583-584

[111] Ojewole JAO, Adesina SK. Isolation, identification and some cardiovascular actions of a purine nucleoside from the roots of Boerhaavia diffusa. Fitoterapia. 1985;**56**(1):31-36

[112] Kadota S, Lami N, Tezuka Y, Kikuchi T. Constituents of the roots of *Boerhaavia diffusa* L. Examination of sterols and structure of new rotenoids,

boeravinones A and B. Chemical and Pharmaceutical Bulletin. 1989;**37**:3214-3220

[113] Kokate CK, Purohit AP, Gokhale SB. Pharmacognosy. 38th ed. Pune: Nirali Prakashan; 2005. pp. 537-538

[114] Aslam M. Asian medicine and its practice in Britain. In: Evans WC, editor. Pharmacognosy. London: Saunders Company Ltd; 1996. pp. 499-500

[115] Jadhav VM, Thorat RM, Kadam VJ, Salaskar KP. Chemical composition, pharmacological activities of *Eclipta alba*. Journal of Pharmacy Research. 2009;**2**(8):1129-1231

[116] Brinkhaus B, Lindner M, Schuppan D, Hahn EG. Chemical pharmacological and clinical profile of the East Asian medical plant *Centella asiatica*: Review article. Phytomedicine. 2000;**7**(5):427-448

[117] Khanna AK, Rizvi F, Chander R. Lipid lowering activity of *Phyllanthus niruri* in hyperlipemic rats. Journal of Ethnopharmacology. 2002;**82**(1):19-22

[118] Suhaj M. Spice antioxidants isolation and their antiradical activity: A review. Journal of Food Composition and Analysis. 2006;**19**:531-537

[119] Gupta VK, Shama SK. Plants as natural antioxidants. Natural Product Radiance. 2006;**5**(4):326-334

[120] JVV D, Jha OP, Mishra A. Chemotaxonomy of Amarathacea. Study of triterpenes. Plant. Biochemical Journal. 1977;**4**(1):14-18

[121] Kapundu R, Lami M, Delande N. Analysis of Saponin from *Alternanthera sessilis*. Bulletin de la Société Royale des Sciences de Liège. 1986;**55**(5-6):605-666

[122] Gamble JS. Flora of the presidency of Madras. Rubiaceae to Euphorbiaceae. Vol. II. London: Adlard & Son, Limited; 1921